PRO REGE ET LEGE

Matthew Murray

Pioneer Engineer

Records from 1705 to 1826

Edited by

E. Kilburn Scott, A.M.Inst.C.E.

TEE Publishing
Warwickshire, England

First published 1928
Reprinted 1999
© TEE Publishing
ISBN 1 85761 111 X

FOREWORD.

THE first part of this volume contains a paper by Mr. G. F. Tyas; also letters of Matthew Murray, M. R. Boulton and James Watt, Junior, reprinted by permission of the Newcomen Society.

The second part contains abstracts from various publications; the Centenary Sermon by the Vicar of Holbeck; a portion of my lecture notes and the various connecting matter by which I have tried to give a complete pen-picture of the engineer and his associations.

For quick reference, two Chronologies are included, which may be helpful to those who have not time to read the whole book. Any who wish to delve further into the subject will find the Bibliography useful, and it serves also to show the extent of the references.

The book is part of a Memorial to Matthew Murray, and surplus income from its publication will go towards carrying out other parts of the Scheme, including the erection of bronze Memorial Tables—

(a) On the site of the " Round Foundry," Holbeck.

(b) In a central position in the City of Leeds.

(c) At the Institution of Mechanical Engineers, London.

It is also proposed to establish annual " Matthew Murray Prizes " for *Day* and *Evening* students of the Leeds Technical College, who show proficiency in mechanical drawing, etc.

This is a very suitable part of the Memorial, because engines built by Murray are described by some writers, as being well proportioned and of elegant design. The awards are to be made on the design work and note books.

If sufficient money is raised, copies of the Matthew Murray models in the Science Museum will be made for the Leeds Museum. The importance of Murray in engineering is shown by the fact that these special models have been given a prominent place in the Government Collection at South Kensington. The Authorities have promised to lend drawings, etc., for similar models for the Museum in Leeds.

The need for a Memorial was mentioned about forty years ago in a lecture I gave in South Leeds, entitled " Leeds Engineers.'' For this, Messrs. Joseph Craven and George March assisted, the latter lending the original picture from which the frontispiece has been made.

In 1912, on the Centenary of the starting of Murray's locomotive, Mr. Walter Fourness and other railway men walked along the track and held an informal meeting at Belle Isle. It was reported in newspapers at the time.

About twenty-four years ago, when the Watt statue was placed in City Square, there were suggestions that a statue of Matthew Murray would have been much more appropriate, seeing that he started the engineering trade of Leeds.

Since the war there have been letters in newspapers, notably from Messrs. H. Backhouse, W. J. Barker, W. Fourness and others. One letter was signed by members of the Institutions of Civil and of Mechanical Engineers; the Leeds Engineers Association and the Institution of Locomotive Engineers. Lt.-Col. Kitson Clark has also made suggestions.

In the meantime Mr. Rhys Jenkins, a former patent agent, of Leeds, who had been carefully going through papers in the Boulton & Watt collection, at Birmingham, discovered some letters of M. R. Boulton and James Watt, Junior, containing important references to Matthew Murray.

Several of these letters were shown in the 1926 Centenary Exhibit at the Science Museum, South Kensington, and later I arranged for them to be lent to the Exhibition in Leeds during Tercentenary Week.

Whilst there, Councillor J. Arnott, the Lord Mayor, authorised photographic copies to be made for the archives of Leeds, and I also took the opportunity to forward a set to the Engineering Societies Library, in New York City.

These letters of James Watt, Junr., would have been good subjects for the trenchant pen of T. W. H. Crosland, and if they had been known at the time of laying out the City Square it might have had a statue of Murray, instead of Watt.

To emphasise further the importance of having a Memorial it may be mentioned that before the Murray Centenary I suggested an article on Murray, to the Editor of a London engineering paper. He said he had never heard of Matthew Murray !

Publication of the facts in this book is, therefore, timely, in drawing attention to the Leeds engineer, and it will also help to prevent erroneous statements. For the sake of the reputation of their City, Leeds people should support the Memorial enthusiastically.

In writing my own notes and editing those of others, I have been careful not to over-state claims, and in several cases have made qualifications. For example, in referring to the metal planing machine I mention that it is sometimes credited to others.

The firm of Fenton, Murray & Wood was certainly the first to start the regular manufacture of machine tools and export them abroad, and the tradition then set up was carried on, on the same site, by Smith, Beacock & Tannett.

It is now customary to show a certain reverence for places where important engineering events have occurred, as for example, the Soho Foundry, Birmingham, and the Stockton and Darlington Railway. From that point of view, the " Round Foundry " site in Holbeck, and the Hunslet to Middleton Railway should rank high.

I hope readers will appreciate, as I do, that there is much romance in engineering history, and certainly the story unfolded in these pages is an interesting section of Leeds Civic history.

It is a pleasure to acknowledge assistance from Messrs. H. W. Dickinson and E. A. Forward, of the Science Museum; Messrs. F. J. March and G. S. Wainwright, relatives of Matthew Murray; Messrs. W. J. Barker, H. Jennings and G. F. Tyas.

84, Kingsway,　　　　　　　E. KILBURN SCOTT,
　　London, W.C.2.　　　　　　　　　October, 1928.

Mr. Murray.

FROM A WATERCOLOUR DRAWING IN THE POSSESSION OF
MR. FREDERICK J. MARCH, OAKBROOKE HALL, DERBYSHIRE.

PART I.

MATTHEW MURRAY.

A CENTENARY APPRECIATION

BY

G. F. TYAS.

Read at the Science Museum, South Kensington, February 24th, 1926, and reprinted by permission of the Newcomen Society.

The career of an engineer who was well known during the early part of the 19th century, and whose work covered almost the whole range of mechanical engineering, as practised at the time he flourished, is the subject of this paper.

Matthew Murray (*See* Frontispiece) is believed to have been born in Newcastle-on-Tyne in 1765. Little is known of his early life, but he seems to have made the very best use of his opportunities at school. His father, it is said, apprenticed him to a smith—but it seems more probable to a millwright or mechanic—in his native town, where he married before the completion of his term of service. He then went to Stockton-on-Tees, finished his apprenticeship, and worked as a journeyman in a mechanic's shop there. Trade becoming bad, and learning that there might be a chance for him in Leeds, where great efforts were being made to introduce machinery in several of the manufactories, he decided to try his fortunes there.

Taking a few articles and his kit of tools, Murray set off to walk to Leeds, a distance of about 60 miles. The morning after his arrival there he tramped to the water mill in the picturesque valley at Adel, about five miles from Leeds, where John Marshall, who was amongst the first to use machinery in flax manufacture, was in business. He required assistance, and at once set Murray to work, first as a handyman, doing odd jobs about the machinery. Later, finding his abilities useful, Marshall made him his chief mechanic, and for several suggestions and improvements subsequently presented him with £20. Murray, now finding his prospects more hopeful, sent for his wife, and took a cottage on Black Moor, near Adel.

After a time the business began to improve rapidly, and the old water mill became too small for the work, so that Marshall, in 1790, decided to remove to larger premises, nearer the Leeds manufacturing area. In partnership with Mr. Benyon, he started a mill at Holbeck, and this was at first driven by a water-wheel, the water being supplied by a Savery steam engine. About this time Murray took out his first patent (No. 1752), of 1790, which was for "a machine for spinning flax, cotton, silk, etc." In 1793, a steam engine by Boulton and Watt, of 28 horse-power, and driving 900 spindles, superseded the water-wheel. Murray had charge of this and he introduced several minor improvements in its working.

In 1793 he took out patent No. 1971 for "Instruments and Machines for Spinning Fibrous Materials, etc.," It included a carding engine and spinning machine. He introduced "sponge weights" and the process of "wet spinning" of flax, which created a revolution in the trade.

He had been with John Marshall for about twelve years when he was invited to join James Fenton, David Wood, and William Lister, in establishing works for making and repairing machinery and engines, there being no facilities of the kind in Leeds at the time. Mr. Fenton found most of the capital, Murray took charge of the engine building department, and Mr. Wood the construction and design of the machinery. William Lister appears to have been a sleeping partner only, but his name appears in contracts, etc. They first took a

8

workshop in Mill Green, Holbeck, but finding this too small, soon removed to a site near to Camp Field, in Water Lane. Here, amongst other buildings, they constructed a large circular workshop, from which the works were, and are, generally known locally as the " Round Foundry," (*see* Plate 1 and Plan 2 on page 35). Murray devoted almost his entire energies to making improvements in the steam engine, which were afterwards the subject of several patents.

He suggested making engines, simpler in construction by placing the cylinder in a horizontal position, and making connection with the fly-wheel shaft by a rack and pinion arrangement, almost similar to a motion afterwards used in printing machinery. He also invented a method of regulating the draught in the chimney, according to the pressure in the boiler, this being effected by a small steam cylinder, the piston rod of which, by means of a rack and pinion, opened and closed the damper. All these inventions were included in his patent No. 2327, of 1799. It does not appear, however, that any engines, as thus described, were actually brought into use, although the self-acting damper was a success and used for many years.

Having built himself a house close to the works, which still exists, he had it entirely heated by steam, hence it was known locally as "Steam Hall." Here, he afterwards hospitably entertained Mr. Murdock, of Messrs. Boulton and Watt, when visiting the town, and it was subsequently the scene of an exciting episode when a party of the " Luddites " visited the various works with a view to stopping the use of machinery. Mr. Murray being from home, his wife, after refusing to parley with the leaders, presented a pistol at them and fired; it does not appear that anyone was injured, but they immediately decamped, and did not trouble the place any more.

In 1801 Murray took out a patent (No. 2531) described as for " New Methods and Improvements of Constructing the Air Pump and Sundry other Parts," this also including drop valves, valve gear, cylinder packing, parallel motion, and a method of constructing boiler fire places for consuming their smoke, etc.

The following year, 1802, Murray, thinking there might be a demand for a more simple and compact design than the beam engine, devised a type of vertical engine with all its parts arranged on a cistern containing the condenser, air pump, etc., and his patent (No. 2632) also included several forms of slide valves, including the well-known three-ported valve. The title of this patent was for " New Combined Steam Engines for Producing a Circular Power, and for certain Machinery thereunto belonging, Applicable to the Drawing of Coals, Ores, and other Minerals, or for any purpose requiring Circular Power." The general arrangement somewhat resembled a table engine in appearance, with connecting rods working downward, but the connection to the fly wheel shaft was very different, being a species of parallel motion which depends on the principle that " a cycloidal curve formed by one circle rolling within another becomes a straight line when the diameter of the outer circle is twice that of the inner one." (See Plate 3.) This form of parallel motion invented by Murray was communicated by him to Mr. James White, who published it in his " New Century of Inventions " in 1801, and who, when residing in France about that time, submitted it to Napoleon Bonaparte, from whom he received a medal. Consequently it is generally known as White's parallel motion.

These self-contained, so-called " portable " engines were amongst the first to have a simple slide valve, which was worked by a wiper or cam. Several engines of this design were constructed, one of which was fixed at St. Peter's Quay on the Tyne, and is said to have worked uncommonly well for many years. The writer has an early recollection of seeing one of these engines lying in a dismantled condition at a disused colliery near Leeds, where it had been used for winding. Murray also brought out another form of vertical engine, but with the shaft over the cylinder, and with the same parallel motion. It is interesting to note that Murray in his patent specification particularly states that the fly wheel has a rim of circular section, which he " considers a part of this invention."

In the following year, 1803, Murray designed and constructed several beam engines of fairly large size, having columns for supporting the main centre bearing of the beams, instead of placing them on a wall running across the engine house. One of these engines was sent to a large iron works in Staffordshire, where it worked for many years.

About this time Murray's firm threatened to become a serious rival to Messrs. Boulton and Watt, who hitherto had almost a monopoly of the steam engine trade. They sent William Murdock and another of their engineers to Leeds, where they called on, and were cordially received by, Mr. Murray, who showed them round the works. After they had returned to Birmingham, Messrs. Boulton and Watt almost immediately took steps to contest the validity of Murray's last two patents, which they managed to get repealed, or set aside, on the grounds " that Murdock had in one case proposed using similar circular drop valves," although he had not patented them, and in the case of the patent for the portable vertical engine " that the condenser and air pump were copies of Watt's arrangement." The coincidence between the inventions gave rise to a lawsuit, in which the opinion of the Court was given against the patentee (Murray), and although the other schemes which were described, and which all allowed to be original and unquestioned, were at the same time declared void or rather destroyed or lost to the inventor, in accordance with " that most extraordinary principle of the Equitable Law, that a patent must be good for everything contained in it or good for nothing." Thus, by including too many improvements in each of his patent specifications, instead of taking out separate patents for each of them, Murray lost the whole.

He was naturally enough very much hurt at losing the fruits of his labours, and from this time, and for many years afterwards, there was very bad feeling and strong rivalry between the two firms. Each strove by every means in their power to excel in design and workmanship, and this tended to greatly advance their respective productions.

Murray issued an advertisement, and also a challenge to Messrs. Boulton and Watt, which appeared in the *Leeds Mercury* of 20th July, 1803, and various other newspapers, and of which the following is a copy :—

MATTHEW MURRAY v. BOULTON & WATT.

Patent Steam Engine Manufactory,
Leeds.

" I feel myself called upon to vindicate my character as an Engineer, against a foul insinuation in a Paragraph inserted in the Newspapers of last week, I suspect by Messrs. Boulton, Watt & Co. ; they assert ' that every Improvement which was really new and useful, and deserving a Patent, in the one which I obtained in the year 1801, was invented and practised at their Works, and that I surreptitiously obtained a knowledge of them from some of their Workmen.'

" I do positively deny that I ever got the hint of the Improvements in Question from anyone, indeed a little observation is sufficient to refute their assertions if I knew that these Improvements had been invented and practised at Soho, I must have been deficient in Common Sense as well as Honesty to attempt to obtain a patent for what I knew I could not hold. Had they used my inventions in the manner described in that Patent prior to the date thereof, they certainly would have practised them in the Engines they made before that period, or taken out a Patent for the Improvements themselves.

The reason of my not defending the Patent was not from fear of losing the Trial as they seem to insinuate, but that I did not think proper to defend it with such rich and powerful Opponents as Messrs. Boulton, Watt & Co. ;—but had I been guilty of obtaining a Knowledge of their Improvements, if they had any, (but I do not believe they had made any worth Notice since Mr. Watt senior retired from the management), it would only have been a return in kind ; Mr. Storey, Manager of their Foundry, and Mr. Murdock, Superintendent of the Workmen at Soho, some time back visited our Works at Leeds, and from their assuring us of Messrs. Boulton, Watt & Co.'s friendly disposition were admitted into every part of the Manufactory by Mr. Wood and myself ; they were permitted to take Patterns and Specimens of our Workmanship, and we know that upon their return to Soho many of our Improvements were immediately adopted, and the engines made after that by them were in part constructed on our Plans.

" Mr. Murdock, upon taking his leave of us, expressed a wish that as they and we were certainly the best Engine Makers in the Kingdom, we should always be upon good terms, and that if ever I should go to Soho, they would be very glad to show me all their Works.

" I did go to Soho, and was refused admittance into their Manufactory of Steam Engines

" But the World I believe cares very little about Messrs. Boulton and Watt stealing my Inventions, or my stealing theirs ; what people want of us are good engines, but I am confident I can make good ones ; and as they hint that no one can do that but themselves I am willing to end this dispute for the good of the Public in a similar method to one they proposed to Mr. Hornblower with whom they had a dispute some years ago, when Mr. Wilson, their Agent in Cornwall, gave him a Challenge for £1,000 that Messrs. Boulton and Watt would produce an Engine superior to that Mr. Hornblower had created at Tincroft Mine. This Challenge Mr. Hornblower did not accept. Now, I offer by way of Trial, and Proof of Ingenuity and Workmanship, to make an Engine against one of the same Power, to be contrived and made by Messrs. Boulton, Watt & Co., and I do further offer to deposit in the Hands of any Banker in London if they will do the same, One Hundred Guineas ; to become the Property of the Party whose Engine is declared to be most perfect and useful by twelve practical men, (Practical Engine Makers), six to be chosen by Boulton, Watt & Co. and six by me.   MATTHEW MURRAY."

Messrs. Boulton and Watt purchased the vacant land adjacent to Murray's works, and retained possession of it for a number of years, thus preventing any extension of them. (*See* Appendix B.)

Instead of being discouraged by these rebuffs, Murray now applied himself with increased energy to the business, orders flowed in, and in the next year, 1804, they commenced sending work abroad, an engine for Sweden being amongst the first of their foreign orders. Mill work and the making of machine tools were now undertaken, several large boring machines and a machine for planing valves and similar work being made.

In 1806 Murray designed and built an engine with the beam placed below the condensing cistern and cylinder, the connecting rod working upwards to the crank shaft. This reduced the height of the engine, but, as the working parts were found rather inaccessible, only a few of these engines were made. (*See* Plate 4.)

During the next year, 1807, several beam engines of fairly large size were turned out, two of them going to water works in London and others to various places in the Midlands. In 1809 Murray invented and brought out a flax heckling machine, and for it he received the gold medal of the Society of Arts, which was presented to him by the Duke of Sussex. Murray again tried further to simplify and improve the beam engine, and in 1810 brought out a type that was almost self-contained, all the working parts being assembled on a cast-iron plate with the condenser and pumps attached to the under side, only the outer bearing for the crank shaft being separate. Four inclined columns supported the plummer blocks for the beam, and an ordinary three-ported slide valve was used, this having a rack on its back and a toothed quadrant working it by means of an eccentric. The writer remembers several of these engines at work, one at Murray's works, another of a larger size driving a woollen manufactory, St. Peter's Mills, Burley Road, Leeds, whilst a third was in an oil mill at York. (*See* Plate 5.)

In 1811, Murray appears to have been working in conjunction with Trevithick, for we find that he constructed one of his high-pressure engines, which was fitted in a boat the following year, 1812. The vessel was originally a French privateer lugger, " L'Actif," which was captured, brought to Yarmouth, and bought by a gentleman named Wright, who proposed using it for passenger service between Yarmouth and Norwich. The engine had a cylinder 8 in. diameter and 2 ft. 6 in. stroke, and the boiler had a wrought-iron barrel with cast-iron ends. The vessel was brought round to Leeds under sail, to the Canal Basin, which was close to Fenton, Murray & Wood's Works, and the engine, boiler and paddles fitted. The boat was then " steamed " to Yarmouth, and when put into regular work often ran at a speed of from ten to twelve miles an hour, and proved a great success. This was a very early application of steam to regular passenger traffic in this country.

About the same period, Mr. John Blenkinsop, who was the manager of Brandling's Colliery at Middleton, near Leeds, patented a system of accomplishing the haulage of the coal wagons on the colliery railway there. The line, which communicated with Leeds, 3½ miles distant, was in three sections, two of them being comparatively level with a rather steep incline between. This was afterwards converted to a self-acting incline, the loaded wagons coming down drawing up the empty wagons. Blenkinsop's Patent (No. 3431) of 1811, is worded, " For certain mechanical means, by which the conveyance of coals, minerals and other articles are facilitated, and the expense is rendered less than hitherto." Beyond saying that " I do declare that a Steam Engine is greatly to be preferred," an engine is not described, but only the rack-rail and cog-wheels.

14

It was Matthew Murray who designed and constructed the engine for Mr. Blenkinsop. (*See* Plate 6.) It had two vertical double-acting cylinders, partly sunk into the boiler, and very similar to Trevithick's arrangement, but each driving on to an intermediate shaft, with pinions working a cog-wheel which engaged with the rack rail on one side of the track. The two shafts were so arranged that the cranks were at right angles, thus ensuring certainty in starting, besides dispensing with a fly-wheel. The boiler contained a single flue, and it evaporated 8 cubic feet of water an hour with a consumption of 75lb. of coal. Four-way cocks were used for the steam distribution, again following Trevithick's practice, although some years previously Murray had invented several forms of slide valves, including the box or three-ported valve that Murray introduced in 1802. The engine weighed 5 tons and cost £400. Three other engines were afterwards supplied to the Middleton line and one to the Coxlodge Railway on the Tyne near Newcastle. The engines ran for many years, one until 1835, and seem to have had few breakdowns.

The writer recollects a good representation of one of these engines on the sign-board of an inn, " The Old Engine," near the foot of the self-acting incline on Hunslet Moor, Leeds. This was the first railway for which an Act of Parliament was obtained, the first rack railway, and it was also the first railway on which any steam locomotive was a *commercial* success.

The railway was visited by many people, including eminent engineers and scientific men. Mr. J. U. Rastrick and Mr. J. Walker made separate reports about it for the Directors of the Liverpool and Manchester Railway. It was long one of the sights of the district and did a great deal towards making Leeds a centre for railway material.

The following account of the first public trial of the engine appeared in the *Leeds Mercury* of June 27th, 1812 :—

" On Wednesday last a highly interesting experiment was made with a Machine constructed by Messrs. Fenton, Murray and Wood of this place under the direction of Mr. John Blenkinsop, the Patentee, for the purpose of substituting the agency of steam in the conveyance of coal on the Iron Railway from the mines of I. C. Brandling, Esq., at Middleton, to Leeds.

"This machine is in fact a steam engine of four horse power, which, with the assistance of cranks turning a cog-wheel and iron cogs placed at one side of the Railway, is capable of moving when lightly loaded, at the speed of ten miles an hour.

"At four o'clock in the afternoon, the Machine ran from the Coal Staith to the top of Hunslet Moor, where six, and afterwards eight waggons of coals, each weighing $3\frac{1}{4}$ tons, were hooked to the back part. With this immense weight, to which, as it approached the town, was superadded about fifty of the spectators mounted upon the waggons, it set off on its return to the Coal Staith, and performed the journey, a distance of about a mile and a half, principally on a dead level, in twenty-three minutes without the slightest accident.

"The experiment, which was witnessed by thousands of spectators, was crowned with complete success, and when it is considered that this invention is applicable to all Railroads, and that at the works of Mr. Brandling alone the use of fifty horses will be dispensed with, we cannot forbear to hail the invention as of vast public utility, and to rank the inventor amongst the benefactors of the country."

On another occasion the engine took a train of thirty loaded coal wagons at a speed of $3\frac{1}{4}$ miles an hour, and this was witnessed by the Grand Duke Nicholas of Russia and other eminent foreigners, whilst visiting the manufacturing districts of this country. Plate 7 shows a part of the line.

In 1813 Murray designed and constructed some large beam engines, several of which had a spur wheel on the crank shaft driving, by means of a pinion, a second motion shaft on which the fly-wheel was fixed. An engine was built and put to work in that year at the Water Hall Mill, Holbeck, which the writer recollects very well, and particularly its three wagon boilers. This engine worked until 1885, a period of seventy-two years.

Murray and his firm gave much attention to improving their designs, making their engines more symmetrical and elegant in appearance, and they now brought out a type with standards instead of columns, of which many were put to work in the Yorkshire factories, and several in London. The Figure on next page shows standards for supporting the beam and Plate 8 shows single columns.

Murray devoted attention to machine tools, textile machinery, plant for dye-works and warehouses, etc., and he brought out and patented a form of hydraulic press for packing or baling cloth, etc. (No. 3792, of 1814). It was very quick in action, as both tables were made to rise and fall

North East view of Messrs. Fenton Murray & Woods Iron Works at Leeds 1816.

PLATE 1.

VIEW OF WORKS OF FENTON, MURRAY & WOOD, IN HOLBECK, LEEDS, 1806.
The round building on the left is the Fitting-up Shop, which gave
the name " Round Foundry " to the premises.

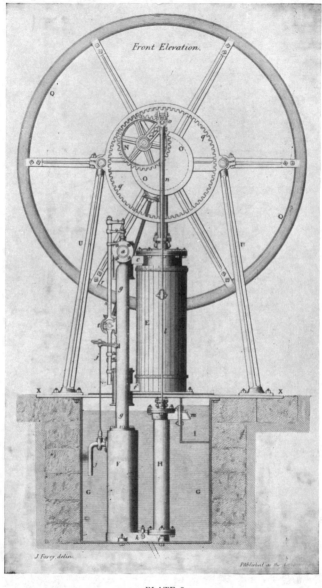

Front Elevation.

PLATE 3.

MURRAY'S ENGINE WITH CYCLOIDAL STRAIGHT-LINE MOTION, 1802.
(From Farey's " Steam Engine," 1827.)

simultaneously, this being effected by racks and pinions. He was the first to apply an indicator or pressure gauge to the hydraulic cylinder, as described in his patent.

In 1815-16 Mr. Francis B. Ogden, who was Consul for the United States in Liverpool, and afterwards engaged in various business projects in America, came to this country, and ordered from Fenton, Murray and Wood several engines for steam boats, one of which was sent out and fitted on board a steam tug on the River Mississippi. (*See* Figure below.) This was a double beam engine to Murray's design, with cranks set at right angles and driving the paddle shaft by spur gearing. It is said to have frequently taken against the stream a ship from 300 to 400 tons on each side, and two smaller vessels, brigs or schooners, towed astern. (*See Mechanics Magazine*, 1830, p. 147.

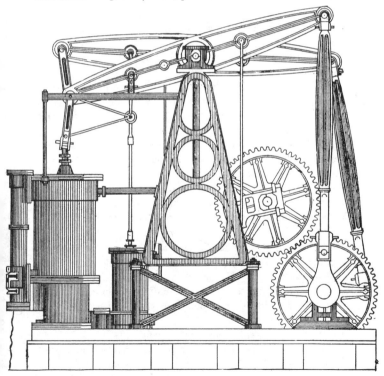

Fenton, Murray & Wood were pioneers in the supply of engines, machinery, and mill-work for factories on the Continent. The Emperor of Russia, in recognition of Matthew Murray's services, presented him with a valuable diamond ring; and he also received a gold snuff-box from the King of Sweden.

In his later years, Murray was engaged in almost every class of engineering work, including drainage of marsh lands, and gas and water works. He was the first in Leeds to have his works lit with coal gas, and he was largely instrumental in inducing the authorities to consent to the construction of works for supplying the town with gas. He also acted as a consulting engineer to woollen mills, dye houses, oil mills and coal mines.

He died on February 20th, 1826, in his sixty-first year, and the *Leeds Mercury,* in announcing his decease, says of him, that he was "A man whose mechanical abilities were, perhaps, inferior to none, his great improvements in the steam engine, flax spinning and other machinery will be a lasting testimony of his unceasing labour." He was interred in a vault in Holbeck Churchyard, Leeds, where a cast-iron obelisk marks his resting place. (*See* Plate 9). He was of a genial temperament, with a frank, open disposition, and free and accessible to his workmen. He had an intimate perception of design, was an excellent draughtsman, and his designs were so well-proportioned that it was very rarely that any alterations were required.

He left three daughters, two of whom married the partners of the well-known firm of Maclea and March, Machine Tool Makers, Union Foundry, Dewsbury Road, Leeds.

After his death, the firm of Fenton, Murray and Jackson continued the business for some considerable time, and, among other work, built some marine engines for the post packet service between France and Constantinople. These were of the side-lever type, and their design and fine finish were greatly admired by a number of visitors before the engines were sent away for shipment.

The firm then turned their attention to the building of locomotives, and supplied a large number to various railways, including the Liverpool and Manchester; Leeds and Selby; North Midland; London and Southampton; Paris and Versailles; and Great Western Railways. Most of these had to be sent to their destination by canal, and in shipping one of the Great Western engines, it fell into the water, much trouble being entailed in recovering it. When the Great Western engines were finished, a testing machine with "live" rollers was built for their trials, this being probably the first instance where such a contrivance was used.

To the uninitiated it may be explained that " live " rollers are a mechanical device to test the power of a locomotive engine without its moving forward ! its track wheels rotate the " live " rollers.

The famous " Round Foundry " Works were noted for the early training of many celebrated engineers, among whom may be mentioned :—

Benjamin Hick, who afterwards founded the firm of Hick, Hargreaves & Co., Bolton, Lancashire.

Richard Peacock, of Beyer, Peacock & Co., Gorton Foundry, Manchester.

Luke Longbottom, Locomotive Superintendent, North Staffordshire Railway.

Murray Jackson, chief engineer of the Imperial & Royal Danube Steam Navigation Co.

David Joy, the well-known inventor of Joy's valve gear, Locomotive Superintendent of several railways.

J. Chester Craven, first Locomotive Superintendent of the Brighton Railway.

Fenton, Murray and Jackson closed the works in 1843, and the works were eventually taken over by Messrs. Smith, Beacock and Tannett, machine tool makers, who renamed the premises " Victoria Foundry," and occupied them until 1894. Several engineering and other firms now occupy the premises. By some Holbeck people they are still known as the " Round Foundry."

The writer is indebted to the Director of the Science Museum for access to the Murray letters in the Goodrich Collection, and to Mr. Rhys Jenkins for the letters relating to Murray in the Boulton and Watt Collection. The most important of these are printed in Appendices A and B.

The paper was illustrated by a number of drawings and photographs, from which Plates 1 to 9 have been prepared.

Before commencing the meeting, the Special Murray Centenary Exhibition in the Museum was inspected.

DISCUSSION.

MR. E. KILBURN SCOTT, in opening the discussion, referred to the striking memorial service for Matthew Murray which was held in Holbeck Church, Leeds, on the previous Sunday, when an appropriate sermon was preached by the Rev. R. J. Wood, and the Lord Mayor, Councillor John Arnott, and other leading citizens were present.

Dr. Samuel Smiles, who lived in Holbeck and who was the Editor of the *Leeds Times,* wrote about Murray in *Industrial Biography* and other publications. The Rev. R. V. Taylor, Vicar of St. Barnabas, Holbeck, also wrote a book entitled *Leeds Worthies,* in which Murray is mentioned. Footnotes to this are contributed by Mr. J. Ogdin March, a son-in-law of Matthew Murray. Murray was in all probability apprenticed to a machine smith or millwright, the forerunners of present-day mechanical engineers. Evidence that he was a good smith is afforded by the fact that he gave a fine example of his work to Murdock when he visited Holbeck. It was generally said that Murray had little education, but he was an excellent craftsman, which is a very desirable form of education. His letters, and the calculations contained in some of them, showed that he was educated in a book sense also. Then he married the right woman, and Mrs. Murray's character is portrayed by her action in the " Luddite " episode. He came into contact with John Marshall, who was a man of liberal ideas and wanted to use machinery to do work previously done by hand. The first mill of Marshall and Benyon in Holbeck was driven by an overshot waterwheel, operated by water that was pumped from the adjoining beck by a Savery engine. Murray had charge of this engine and he thus had an excellent opportunity to learn about steam power.

A series of slides was then shown by Mr. Kilburn Scott, comprising :—Portrait of Matthew Murray; View and plan of the Round Foundry premises as shown in Plates 1 and 2 of this paper; Heckling machine for use in flax spinning. This was the machine for which Murray was awarded a Gold Medal by the Society of Arts; Engine with straight line motion; Engine built for Francis B. Ogden, of U.S.A.; Side lever engine; Locomotive of 1812—Murray carried out experiments on a locomotive at his works, and he had built one by 1811; Yorkshire collier of the period; Rack rail six feet long; Diagram of engine with rack-rail and cogwheels; Articulated locomotive, which showed that Murray knew about engines running on smooth rails; " Steam Hall " or Holbeck Lodge; Monument in Holbeck Churchyard, the inscription on which refers to Murray as a Civil Engineer.

There was no doubt that Murray was largely responsible for establishing the flax-spinning industry in Leeds, and he revolutionised the trade by inventing " wet spinning " of flax. He is said to have made the first metal planing machine for machining D valves. He made the first boring mills with screw feed. Altogether he was a man of wonderful character, and the different episodes in his life afford an interesting study to-day by others as well as engineers.

MR. E. A. FORWARD remarked that it was curious that the Science Museum appeared to be the only place that contained any documentary evidence concerning Murray. He read one of Murray's letters, which was a reply to a request from Simon Goodrich for knowledge upon steam engine matters in general. (*See* Appendix A.) With regard to the rack rail, Farey's account states that he was at Leeds when Murray was experimenting, and that the rails were then 3 feet long with hollow teeth. Solid teeth probably came after the hollow toothed short rails. There seemed to be no doubt that Murray made experiments before he constructed the locomotive for Blenkinsop. He was somewhat afraid of engines using high pressure steam, which explained the close resemblance between his own and some engines of Trevithick. The introduction of the cylinder boring machine with screw feed is also attributable to Murray. The Watt letters show that

Murray received some shabby treatment from Boulton and Watt. (*See* Appendix B.)

Lt.-Col. E. Kitson Clark, in a letter read by the Hon. Secretary, said that Matthew Murray was a moderate man as well as a genius, and he felt the necessity of simplifying mechanism as far as possible; this idea pointed out to him the desirability of bringing together the inlet and outlet openings in the valve box, hence his patent.

His use of the reciprocating engine for textile purposes was distinctly a bold move, and there must have been a picturesque side of his nature, which showed itself when he erected his Round Foundry. The shape of this building would have required a polygon of shafting, and this would have necessitated the use of gears.

As a member of a firm of locomotive builders in Leeds, the writer expressed his strong sense of indebtedness to the great pioneer in locomotive work, and his sense of the honour that he carried out his experiments, gained his experience, and achieved success in that city.

Mr. H. W. Dickinson said that Mr. Forward had referred to the fact that the Science Museum possessed all the matter of technical value connected with Murray. Murray must have been a pioneer in the development of hydraulic machinery. In one specification a pressure gauge, in which one cylinder acted upon another cylinder of diminished diameter, is mentioned. A number of references had been found by the President in the Boulton and Watt Collection.

Mr. Pendred asked if there were any direct evidence that polygonal shafting was used at the Round Foundry, as suggested by Colonel Kitson Clark's letter.

Mr. Kilburn Scott said that the round building was burnt down in 1872, and thus Murray's patterns, drawings, and papers were unfortunately lost. Regarding shafting, Murray would have no difficulty in arranging it in any way. He was an expert millwright and here were men already in the district with the special knowledge of making all types of gear wheels and doing millwright work.

Mr. Forward said that James Watt, Junr., described the Foundry, in 1802, as being built on three floors. The engine was to be placed in the centre, and this would suggest radial shafting.

The President, Mr. Rhys Jenkins, remarked upon the strong feeling of veneration for Murray that existed in and around Leeds. Without doubt, he was the father of mechanical engineering in Leeds. He would like to have heard more about Murray's personality, and thought that in character he seemed somewhat like William Murdock. Murray completed his apprenticeship at Stockton. Now Darlington was the place where flax-spinning was done by machinery, and as there were not many machinery shops in Darlington, it seemed very probable that the Stockton people supplied the machines for Darlington. When flax-spinning by machinery was introduced into Leeds, the Stockton people went there also to fit up the machines, and Murray did a lot of work on flax-spinning machinery. Mr. Tyas had said that the cycloidal parallel motion was due to Murray. What evidence was there to show that this was so? According to Farey, it seemed that the Round Foundry soon had a number of competitors.

Captain E. C. Smith said that it was regrettable that the boiler supplied by Murray for the steam boat at Yarmouth should have burst.

Mr. Kilburn Scott remarked that when it burst it was not in the same state as when delivered by Murray. It had been repaired and got furred up by salt water, etc., and this was the real cause of its failure.

Mr. Tyas replied that with regard to the cycloidal parallel motion, the *Mechanics' Magazine* mentioned that it was made by Murray and communicated to James White. He had not been able to discover any other information concerning it.

Mr. Pendred proposed a vote of thanks to the Director of the Science Museum for the use of the meeting room. This was seconded by Mr. Legros and carried unanimously.

APPENDIX A.

LETTERS FROM MATTHEW MURRAY TO SIMON GOODRICH, ENGINEER AND MECHANIST TO THE NAVY BOARD.

(Preserved in the Goodrich Collection in the Science Museum.)

Leeds Jany 3rd 1804.

Mr. Goodrick.

Sir,—We received your Favor of the 31st Ult. & can inform you that we have by us an Engine compleatly fitted up of 2 Horses power, adapted for pumping water & made upon the most simple Principles, by very little attention an entire Stranger may be learnt to manage it, who might be instructed while it is putting up by the Person we send for that purpose. This Engine is completely fitted up in a Cast Iron Cistern, & needs no other fixing than bolting upon the Foundation it is to stand on & requires no Wood work whatever, the Boiler to be set in brick as in all other Cases. The whole weight will be nearly 2 Tons, and may be conveyed very well by Land Carriage at about £8 P.Ton, or by Water at £2. This Engine is capable of pumping with a 12 Inch Pump 120 Gallons of Water 20 Ft. high Pr. Minute, and more or less in proportion to the height required. The price of this Engine including the Boiler, & Door Frame & Grate Barrs for Fire Place is £175, deliver'd at Leeds & payable as soon as put up, we provide an experienced Engineer to put it up whose Wages, Board & Travelling expenses will be charged to you, should the above meet your Approbation should be happy to attend to your early Order or give any additional Information you may require.

I am for Fenton & Co.

Your mot. Obt. Sert. MATTHEW MURRAY.

Leeds 4th March 1813.

Dear Sir,—I have your obliging favour & shall be happy to give you every Information in my Power—With respect to the Corn Mill the grinding of 6 Bushels per Hour on each pair of Stones, 4 feet Diameter is correct, or a 20 Horse Engine will grind & dress 20 Bushels per Hour—We have a 6 Horse Engine working hear which drives one pair of Stones with dressing Machinery and grinds full 6 Bushels per hour. I think the time you have limited the Mill to work is too much (viz) 142 Hours per Week allowing very little Time indeed for cleaning & repairing; Would it not be better to have a 30 Horse Engine with 5 pair of Stones, One to stand & four to work, this is about the proportion of Power at our best Mills in this Country, and will allow ample Time for all kinds of casualties that might happen.

24

On the other side are the Prices of the Metallic Materials of our Engines exclusive of the Boilers which is specified in the next Column (delivered here). Engines above 12 Horse Power are made with what is called D Valves on a very improved Construction.

Multiplying Wheels are now seldom used, but a heavy Fly Wheel substituted in their place which is much better.

On the other side is the Sketch of a Ground Plan & Elevation of a 20 or 30 Horse Engine House, the black figures are the size of the 30 Horse & the red figures the size of the 20 Horse—Engines that work with Salt Water, we do not use of condencing Cisterns, nor Cold Water Pumps, if the bottom of the Engine House can be placed within 3 or 4 Yards of the Surface of the Water, the Water will then be raised into the Condencer by the pressure of the Atmosphere, the advantage of this method is that it keeps everything clean & good to get too. I need not mention to you that in this case the nearer the Water is to the bottom of the Engine House the better. The Plan you propose for the Copper Boilers will answer exceedingly well & in a great degree prevent the expansion from injuring the brick work. You mention 2 Boilers, I beg to observe that the mode which is adopted now is to have 3 small Boilers, 2 to work together equal to the Power of the Engine (instead of a single Boiler) & the other to change upon, the advantages are their being more portable, consequently easier to replace, also taking less Coal & keeping the Steam more regular another advantage is, that when only one or two pair of Stones are used one Boiler will do. The Quantity of Coals necessary for working these Engines will be for a 20 Horse about two Bushels & a half per hour, for the 30 Horse about 3 Bushels & a half per hour of good Newcastle Coals & in general will grind about 9 Bushels of Wheat with one Bushel of Coals. It has been found that when the Stones are sharp & in good Order & the Wheat Dry & great Economy used in the Coals one bushel of Coals have ground 10 Bushels of Wheat, but in ordinary Cases about 8 Do —Copper Boilers will cost four & a half Times the Price of Best Iron Boilers as you will perceive by the List—When you have determined upon your Size & Method we should be glad to hear from you or give you any additional Information you may require.

Waiting the honor of your further Commands

I remain For Fenton, Self & Wood

Your most obedient Servant   M.MURRAY.

NOTE BY EDITOR.—The Prices above-mentioned are given in tabular form on the next page, and it is interesting to see that the D valves which Matthew Murray had introduced were used on most of the engines.

The Sketch of Ground Plan and Elevation mentioned above, is omitted from this reprint.

Horse Power	Price of Engine Materials	Price of Single Iron Boilers	Price of Single Copper Boilers	Length of Stroke		Strokes per Minute	Additional cost e.g Brass Air Pump Bucket Copper Rod &c.	Injection water required per minute
	£	£	£	ft.	in.		£	
6	400	40		3	0	36		
8	468	60		3	6	32		
10	557	70	315					
12	644	82		4	0	28		
14	716	100						
16	770	100		4	6	25		
18	827	130						
20	894	140	630				140	60 galls
22	963	150		5	0	22		
24	1028	160						
26	1083	175						
28	1134	190						
30	1186	200	900				200	90 galls
32	1228	215		6	0	19		
34	1266	230						
36	1304	240						

Leeds Dec. 3rd 1814.

Dear Sir,—I have your esteemed favour this morning and am very happy to hear that you feel comfortable, from the short Time of your absense from them, they perhaps have been able to appreciate your Value, as in some Situations Merit is discovered by its Absence — — — With respect to the Effect of Hydrolic Pressure, I am happy in being able to give you every Information you may require, as I have lately thought a good deal upon the subject & have taken out a Patent, for what I conceive Improvements upon that Principle & am now in the habit of making them daily for different purposes, & have not any doubt but it is the best possible Principle you can apply for the purpose of trying the Strength of large Cordage, particularly where immense Power is required ; as you do not know what Power you may require, I think you should go as far as 1000 Tons & which you may easily do by a Ram of 16 Diam. operated upon by a forcer of ½ Inch Diamr—This may be pressed down by any Power not exceeding a Ton, for the Power is as the relative Diameters, & in this Case would be 1000 to 1.   If you use a Forcer of a larger Diamr. say, one Contain'd 500 Times in your Ram that would be the limit of your Power in that Case, however it will be necessary for you to have 3 different sized Forcers, as their Expense is not very great & all may be fixed in one Cistern it will be necessary for you to have the Application of an Instrument or Instruments that may denote the quantity of Power upon the Ram at any one Time,—this is part of my Patent & to the use of which you shall be every welcome—In considering how the principle may be applied of stretching a Rope, it occurred to me that the Cylinder might lay horizontally, but this I find has many Objections, on account of its being a very unfavourable position for the Leathers, on which the Perfection of the whole depends, on the other side is a design for placing the Cylinder perpendicular, which is the most natural Way, but this will require a Strong Chain & Wheel to change the Direction from the Perpendicular to a horizontal Pull ; it might be fixed on Stone Work, & a little building contrived to inclose the whole. For your better understanding it, I have sent a Box by this Nights Mail containing the Specification of my Patent, which when you have perused you will please return, carefully directed to us, there is also a Section of a full sized forming Pump which you may keep. I thought it would be saving Time with you. or would have sent you a Copy instead of the Specification. I cannot exactly say what it will cost till you have determined upon your Size but Forcers fitted upon Cisterns with Levers &c. will Cost from 25 to 30£ each, the Ram & Cylinder are sold by weight, & in a finished State would be worth about 42/- pr Cwt, the wheel Chain & other Parts you might make yourselves if you thought proper, as no Estimate of them could be made until a drawing was made out. The Part for denoting the Weight or Pressure I have not yet put in Practice but there is no doubt of its answering to a certain degree, perhaps nearer than any other Way. The Proportion of thickness that Cylinders bear to the Ram is ½ the Diameter of the Ram. In the Sketch you will perceive I make use of a Top Chain, but this is only to balance the Chain below, & and in some degree likewise to steady the Ram. A Cylinder & Ram of the above dimensions with a 4 ft. Stroke will weigh about 5 Tons a 12 inch ram in Course will be ½ the Power & proportionable cheaper, any further Information you may require will be gladly given   by yours Sincerely   M. Murray.

Leeds Feby. 9th 1815.

Simon Goodrick Esqr.

Dear Sir,—I have your favour and am much obliged by your kind Offer respecting the Bridlington harbour, and should have been happy to have undertaken it, (under your Directions) had my present Engagements permitted, but as I am situated at present, it is impossible for me to undertake any Thing, unless Mechanical, as I am totally confined to our Manufactory here, and with any part of which, I shall at all Times be very happy to serve you—I do not think Mr Chapman a proper person, as he has had very little experience in that Way that I know of—The Collector might be made a very useful Assistant, if he could be made to pull fair—With respect to your new Machine for stretching Cables I think it quite practicable in the Horizontal position and to insure with Safety the direct Action of the Ram, I should propose something similar to the inclosed Sketch, for any Thing I can see the Ram might make a 6 ft. stroke as well as a 4, however when you set seriously about it, if you will send me your Sketches, I will with Pleasure communicate unto you any thing that occurs unto me.

and in the mean Time remain, For Fenton Self & Wood

Yours Mo. respectfully   M. Murray.

Leeds 17th July, 1815.

Sir,—I hope you will excuse the liberty I take in writing to you on the subject of what is called Steam Boats.   It occurs to me they might be used with great Advantage in his Majesty's Navy if properly introduced, and as you are in a situation where the Introduction might properly come under your department, I beg your attention to the following Remarks I wish to make on that Business, viz.   Might not the Power of a 40 or 60 Horse Engine say with two 20 or 30 Horse Cylinders, be applyed on Board a Gun Brigg to work 4 Powerfull water wheels two at each end of the vessel.   The vessel to be fit up merely for carrying the Engine and its Machinery without any Masts or Rigging, and the Crew only to consist of Pilot and Enginemen, might not this vessel be employed as a moving Power, in taking in and out of Port in all winds, the Men of War ships that may be coming to be refit, or to take out to Sea, or Indeed to bring in at all times any small craft bringing Stores to the Dock yard, that may be prevented getting in otherwise, and as the Steam Engine is so generally useful, might it not be used in a many ways in his Majesty's Fleets.   I do not mean on Board the Fighting Men of War, where it will always be objectionable, but as conductor of Fire ships, and as an auxilary or attendant for various Purposes, which would be easier pointed out by Sea-men, such a vessel might attend the Embarkation and delivery of Stores, ordinants &c., working merely as a labourer by which means a great deal of time might be saved.   As these vessels will be adopted sooner or later in the British navy I should be glad of your Opinion on the subject.   I merely mentioned a Gun vessel as being perhaps the easiest to obtain, to make the Experiment of, if it may be so called, whose effects are certain, I am apprehensive that the Power & motion of the wheels in the water, would be a great preventitive of the vessel being destroyed in a Storm, and might be made the safest vessel that can swim, and from its great Power might be made to keep company with any Fleet of Ships.   We have a Steam Boat for Passengers that works between Hull and Gainsbro' that answers the Purpose exceedingly well, but I strongly recommend having the Engine in a vessel by itself and taking the other vessels or Packets in Tow, the favour of your answer will much Oblige, Your Obt. Svts.

for Fenton Self and Wood,   Mw. Murray.

The Boat I mention was made in Scotland and is but a very indifferent thing.

Mr. Goodrich,

Dear Sir,—On my return I find your favour, which should have been replied to much earlier, but have been a month from home.

With respect to Mr. Jessop's metallic elastic spiral Packing, I I cannot give you a favourable account of it.

We got one from them, & tried it for about a week, but found during that time we were using considerably more Coal than with the ordinary hemp packing.

We could not make it stop the steam, as it is hardly possible to make the spiral so fine at each termination as to end with nothing, without leaving a space, through which the steam passes.

We intend to try it again when one of Mr. Jessop's Men is coming this road, that he may put it in his own way, & then we shall compare its effects with the common hemp packing.

I have made several different metallic Pistons, which I considered ingenious in their construction, & with the very best of workmanship, but never found them better than the usual mode with Hemp, & indeed always gave them up.

But in Mr. Jessop's piston there seemed something so bewitchingly simple that I thought I would waste a few Pounds in its trial.

At the same time I have always observed that soft metallic surfaces rubbing upon each other wear out very fast, but this is not the case when one of the surfaces is of vegetable matter.

This is particularly observable in what is called D Valves—the flat surface of which is brass, rubbing upon Iron, but the back or circular part is iron, rubbing against a hemp packing—now although the pressure upon each face (back & front) is equal, the brass and iron faces wear out very fast, & the other seems not to wear at all.

This being the case I despair of meeting with a metallic piston that will not wear both itself & the cylinder at the same time,—even though it should be steam tight.

I am at present making a very large Hydraulic Machine for trying Chain cables, for General Wilson, as soon as it is finished I will send you a sett of drawings of it, & also return yours with thanks.

I believe it will be the most powerful ever made as I intend it to lift 400 Tons.

The only improvement which I have lately made in the Steam Engine, is applying two Condensers, one to condense the steam from the upper side of the Piston, & the other from the under side; this allows twice the time for the condenser to cool *in* its vapour, before it receives a second charge—by this means it makes a better vacuum.

We have just finished a 60 Horse Engine in Leeds upon this plan, & the result is very satisfactory.

Wishing you health & happiness, I am, Dear Sir

yours very truly, Mw. MURRAY.

Dear Sir,—I recd. your favour of the 16th, & I observe that you are going to deliver a lecture to a Philol. Society on the Steam Engine.—I am very sorry to inform you that I have not any Model that would be of any use to you or you should have had it with the greatest Pleasure.

The loco-motive Engines that I have made, or heard of, are all moved upon Iron rail ways, that are nearly upon a level. where there is any considerable rise, it requires one of the rails to be cogged, as I believe you have seen at Leeds.

I don't see any objection why they might not be made to go upon a macadamized road, if a motion was given to all the 4 wheels at one time, & made with broad Surfaces.

The objection to locomotive Engines is, that they must be worked with high pressure steam, say from 40 to 60 pounds upon an inch, above the atmosphere, & of course require strong boilers, & other Parts, which makes them a very objectionable weight—those we have already made, including the water they carry, are not less than 6 Tons for the Power of six Horses

I am afraid nothing can be done by Mr. Perkin's scheme without increasing the danger, however a steam carriage would have been the best to have made his experiment upon, as he pretends to get great power, with little room & no danger.

This I cannot understand & in the absence of direct experiment he cannot satisfactorily explain.

There is another, Mr. Brown's Explosive Gas Engine—this is not original, as it was attempted at least 40 years ago although rather different.—The principle was to drop, by drops a bituminous or inflammable substance on a red hot Plate, at the Bottom of a Cylinder, at the same instant an admission of air caused it to explode & force up the Piston, but this also has gone to the " Grave of all the Capulets."

The rotatory, or circular Engine, is also a thing to be desired, but has not yet been brought to any perfection, when compared with the cylinder & Piston Engine.

I have lately made an Engine of 60 Horses Power in which I have 2 separate condensers, one condensing one stroke of the Engine, & the other, the other Stroke, by which means the condensers are kept much cooler, & a much better vacuum is obtained, than is possible by one condenser, where the water & steam are continually rushing in ;—this you will plainly understand without any further observation.

I think Mr Brunton's revolving fire Grate, the best mode of feeding Engines with Coals, & nearly destroys all the smoke, at the same time regulates the quantity of fire, to the quantity of steam required, & also preserves the boilers at least double the time, than by firing the old way.

The quantity of Coal for each horse Power varies from 10 to 15 Pounds, or even 20 Pounds, from different Coal Mines ;—so that there can be no Established rules for Grates, but our proportion for Newcastle coals is 1 Square foot for each horse Power.

Mr. Brunton informed me that he has frequently to alter his Grates, on that acct. Our proportion for boilers, in which there are no tubes, is 20 Cubic feet of water for each horse power, this is a large proportion but we find it the best.

In small cylinders there is more friction than in large ones, of course we allow them a little larger proportion, which is not necessary in larger Engines. Our cylinders are some-what larger than the nominal Power, that they may work with lighter pressure of steam, which is found to be the best, when they are working upon the Crank.

A higher steam is used for Pumping.

We calculate our Crank Engines at 7 lbs. of steam upon every square inch of the Piston.

We make our air Pump, one fourth the contents of the Cylinder, & for single pumping Engines one Eight.

This is the data of what we call a horse power.

Feet per
lbs. minute.   Ratio
150 × 220 Or lifted at that rate = 33000 }
Or 3300 raised 10 high per minute = 33000 } equals 1 Horse power

Or, Any other weight & velocity producing the same ratio.

Dimensions.

Inches
Cylinder 45 inches diameter = 1590 Area of Piston
                    per minute. Lenh. of dbl. stroke,
Piston making 16 strokes × 14 feet = 224 feet per minute
                    lbs.
Pressure of steam 7 on each Sqr. inch of the Piston

Calculation.
            Pressure lbs on the
area of Piston of steam   Piston.   Velocity per minute
   1590   ×   7 lbs = 11130   ×   224 feer

                    Data
      Ratio   of 1 horse   Horse Powers
  =   2493120   ÷   33000   =   75.549

We have not yet been able to get Mr Wilson's proving Machine finished—unfortunately we have had two waster Cylinders, and are now about to try the third.

It will be an immense machine, and is intended to exert a pressure of 1000 Tons. After it is made I intend to try a sett of experiments with it, upon cast & wrought Iron—to determine and compare the strength in different positions.

As soon as it is finished I will send you a drawing of it, together with your Plans, for which I am much obliged.

I am sorry I have not any models of Engines—I had a small Loco-motive Engine, but it was sent several years ago to Russia, or you should have had it.

We are very busy in the Engine-trade at present, but are sorry that the price of Iron has risen so very much lately, as to render Engines considerably dearer than they were some time back; and I am afraid the exportation of machinery, and allowing Mechanics to go abroad will be a disadvantage to trade in general.

Inclosed I send you a sketch of our Portable Engines, which we make as high as 24 Horse Power—they are reduced to as simple a point, as Engines can be, to obtain the full effect of a cylinder & piston Engine.

We think the Beam the best medium between the piston & Crank, it is so very useful and convenient a slave to attach pumps to, or any other motions; and exhibits the Engine in its naked principles. But Engines are made of all forms, witness some made at London, which are cocked up very similar to a Pagoda Temple, and difficult to be understood, besides, what is worse, they are very unmechanical in their operations, and liable to be much oftener out of order, than the plan inclosed; which I think cannot be altered for the better, for this kind of Engine.   I remain Dear Sir

yrs very respectfully,   Mw. Murray.

Note by Editor.—The above reference to an explosive gas engine is of great interest at this present time when internal combustion engines are so much used. The statement " it was attempted at least 40 years ago," shows that the idea is nearly as old as the rotative steam engine with separate condenser. As an explosive they may have had in mind gunpowder or any cheap detonating powder.

Murray's remark about a rotatory or circular engine as something to be desired is also of interest, as it appears to refer to a steam turbine.

Regarding Brunton's revolving fire grate which Murray thought " the best mode of feeding engines with coals and nearly destroys all the smoke," it was the first successful mechanical stoker. Murray had himself taken out a patent for one in 1801.

The reference to Mr. Wilson's proving machine to exert a pressure of 1,000 tons, refers to a cable testing machine worked hydraulically. Although ordered for 400 tons, Murray built it much larger for the purpose, as stated, of carrying out " a sett of experiments."

This was the first occasion when a research on strengths of materials was carried out with such a powerful machine, and it was the forerunner of large hydraulic testing machines of Wicksteed, made by Joshua Buckton & Co. Ltd., and testing machines made by Greenwood & Batley Ltd.

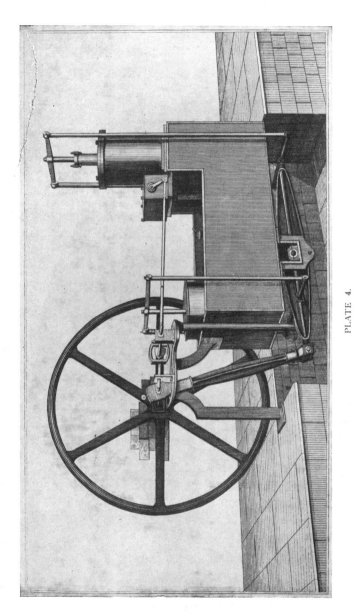

PLATE 4.

MURRAY'S INVERTED BEAM ENGINE, 1805.

(*From Nicholson's Journal of Natural Philosophy, XI., Pl. 7, p. 93.*)

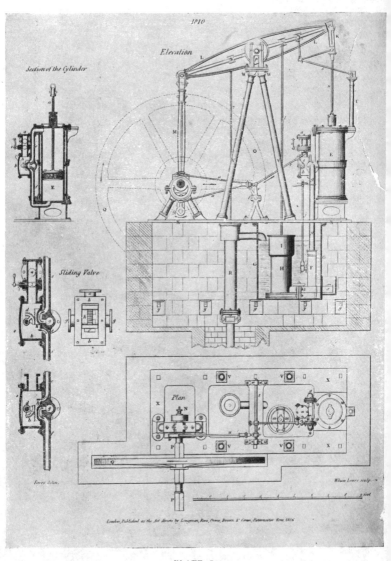

1810

Elevation

Section of the Cylinder

Sliding Valve

Plan

Farey delin.

Wilson Lowry sculp.

London, Published as the Act directs by Longman, Rees, Orme, Brown, & Green, Paternoster Row, 1826.

PLATE 5.

MURRAY'S BEAM ENGINE OF 1810, WITH INCLINED COLUMNS.
(*From Farey's " Steam Engine," published* 1827.)

APPENDIX B.

LETTERS OF JAMES WATT, JUNR., AND MATTHEW ROBINSON BOULTON ABOUT MURRAY'S WORKS.

(Preserved in the Boulton & Watt Collection, Birmingham.)

Letter from M. R. Boulton, Soho, to James Watt, Jnr., London, dated Jan. 17th, 1799.

Murdock & Abraham are now returned from their excursion highly delighted and full of panegyricks upon Murray's excellent work. Abraham is now entirely convinced of his inferiority, and what is more, of the possibility of amendment and he is now *actually* making trials of different substances to mix with the sand with the view of giving a better skin to the castings. We have likewise written to G. Mc Murdock to send a boat load of the sand used by Murray. It seems his forge work is still better than his castings & we are trying therefore to infuse a spirit of amendment into this department of the works. They were admitted into every part of Murray's manufactory & spent two evenings with him and by virtue of a plentiful dose of ale succeeded in extracting from him the arcana any mysteries of his superior performances. . . . Murdock is impatient to commence operations in the boring mill. Everything he saw at Low Moor has tended to confirm him in his opinion of the propriety of separating the different operations. The Low Moor Co., have precisely the same intention with regard to their boring mill which Mr. Dawson means to alter and construct upon a new plan where the different parts will be unconnected.

NOTE.—The Abraham above mentioned was Abraham Storey, the foundry manager at Soho works. G. McMurdock should be McMurdo.

Letter from M. R. Boulton, Soho, to James Watt, Jnr., London, dated Feb. 1st, 1799.

. . . I cannot, however, altogether suppress the pleasure of announcing to you a most promising alteration in the spirit of our forgers & the appearance of their work. I should first inform you that Murray presented Murdock with a specimen of his forge work, no doubt with view of exciting his astonishment and perhaps a despair of ever attempting the same perfection, for I must candidly acknowledge it was the most beautiful and perfect piece of work I ever beheld. Fortunately however it has been the means of producing a very different spirit. The emulation of our men has been awakened and they now vie with each other in adopting the tools and improvements suggested by Murdock. Our first imitative essays have succeeded completely and we entertain great hopes of turning out the working gear in such a manner as will render filing a superfluous operation.

Some other tools, partly in imitation of Murray and partly of the suggestion of Murdock, have been introduced with the quickest success into the fitting department. The new method of turning the larger lathes by means of the endless screw, as shown in the drawings of the boring mill, has been tried upon our air pump lathe & found to answer every expectation. . . . the difficulty of making the endless screws is nearly overcome and a method is now under trial which if it succeeds will enable us to make them very expeditiously.

No improvement, however, in the foundry work—Abraham full of excuses.

———

Holograph letter, James Watt, Junr., to [Matthew] Robinson Boulton, Esqr., Soho, near Birmingham.

Leeds (Sat^s. Evg.) 12 June 1802.

Dear Sir,—We reached this to breakfast this morning and upon a consultation held with Gott determined to employ his millwright, a sagacious fellow of the name of Pritchard to discover the abode of Halligan.   Pritchard occasionally visits Murray's foundry for the purpose of ordering castings & therefor had no difficulty in introducing himself there & singling out Halligan from the description we gave of his person.   Entering into conversation with him he learnt where he lived and returning to us, conducted Murdock to his house, where he found the wife at home & had a conversation of upwards of an hour.   Of the particulars I am not fully possessed but it in all respects confirmed Dixons information. She is exceedingly discontented & her health evidently affected by the state of her mind. She cried bitterly, bewailed her own folly in quitting us, had wished herself a thousand time back among her old acquaintance & wanted to know whether Murdock thought he would agree to take her husband, who she said was disappointed in his expectations though not desirous of going back except upon her account.   That his wages were 22/- per week and he worked at the dry Sand under Dixon. They pay £7 a year for a worse house than they had with us & I believe, rates and taxes.   Provisions &c much about the same price. Young Hughes she said lodged with them, but she had never heard him say anything about his reasons for quitting Soho & knew little about his situation with Murray, as he was remarkably close.— Upon being apprized that I was at Leeds, she expressed a wish that I would permit her to call, which Murdock offered to negociate & desired her to tell her husband that he would be glad to see him at the Inn when he had done work.

I have been surveying the environs of this rival Establishment & making enquiries respecting the property & tenure of the neighbouring lands, with a view to seeing whether we could purchase any thing under their very nose that might materially annoy them & eventually benefit ourselves.   I find there are about 2 acres of Land next field to Murrays works, which may be purchased, but the

34

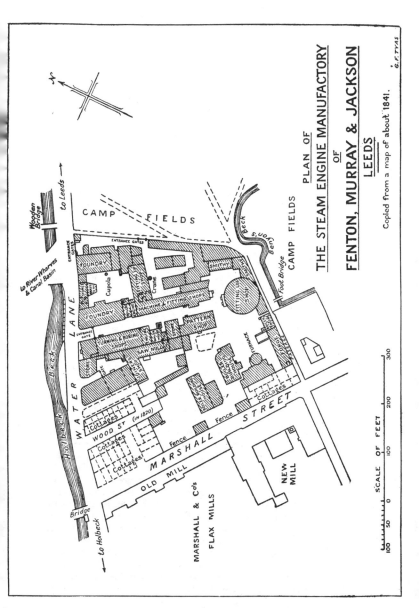

PLAN OF

THE STEAM ENGINE MANUFACTORY

OF

FENTON, MURRAY & JACKSON

LEEDS

Copied from a map of about 1841.

G. F. TYAS

SCALE OF FEET

100 50 0   100   200   300

price probably will be £5 to 600 per acre. I shall learn the exact terms. There is a Malthouse which projects into their premises, which they have in vain endeavoured to purchase at a moderate rate. It is in the possession of a Widow, who is aware that it would be of some advantage to them & therefore asks a high price. This would enable us to overlook their whole Yard & holding it we might dictate our own terms. I shall employ a negociator to learn the terms asked for it. It is said that Murray & Co. have purchased from Marshall the land upon which their works are built, and it is not known that he advances anything to the new *Rotundo,* for such it appeared to me to be, but I shall get a nearer aspect of it to-morrow. Whether the land upon which the Malthouse stands is under lease from Marshall or is Freehold property, I have not yet learnt, but the purchase of the lease would probably answer our purpose. It is generally believed that Murray & Co. are making a great deal of Money & that the new works are constructing from their profits, as also a superb house which Murray is building for himself.—Fenton is, as we formerly heard, a man of some property & was Marshalls partner, who provided him with a share in this business to get him out of his own. Mr. Gott speaks well of him & thinks he would be ashamed of his partners proceedings although he may not have the power to controul him. Wood is the steady man of business who directs the works : Fenton keeps the Cash & Books & Murray solicits orders superintends the erections of Engines &c. He passes him for a great scoundrel but a very able mechanic & Gott says he has got great credit for his last patent, no one doubting that the Inventions are exclusively his own.

Halligan & Pritchard are now here supping with Murdock. I have left them to themselves, conceiving that more information may be obtained in my absence than if I remained with them. I had but a short conversation with Halligan, who has been the greatest part of the evening with Murdock.—He told me that his Wife fretted much & he did not like the country so well as Birmingham, but should have made up his mind to stop here, if she could have been contented. Says he is under no Engagement, tho' most of the other men are, & he has been much pressed to enter into articles. Is employed solely in the dry sand, which he says they do worse than we do & confirms Dixon's report that the nozzles are all made in dry sand. The moulder is dead who made them in green sand and they have no workman now who can do them. Says he does not think much of their moulders although he is sensible of the great perfection of their green sand work, which he attributes entirely to the sand. Says it is exactly the same as we had at the foundry, but that we spoiled it, by mixing too much coal dust with it. The bottom of their green sand foundry (which is a distinct building from the dry sand one) consists of about 4 feet of this sand, which they work over and over again & have never renewed since he has been here (13 weeks). When the mould is formed, a little coal dust is strewed

36

over it, which is all that is used & he supposes this to be consumed by the Iron so that little or none remains mixed with the sand & though its colour is somewhat affected by frequent use, it is not so much changed as might be supposed. Connecting rods, Shafts & Wheels of every description are made solely in green sand, but no pipes or hollow goods. Boxes, where any are used, are of Wood & patched together for the occasion. They melt a great deal from the Cupola, which is about 8 feet high & 20 Inches Diameter & in which they do not use one third of the Coaks which we do to the same quantity of Iron. Believes the Coaks come from near Low Moor. Says the Iron runs much finer from the Cupola than ours does & about 6 Cwt per hour. The worst burnt stuff they can get comes out excellent. Cylr. bottoms, tops & pistons are all made in green sand, until they come to be very large. They had a terrible blow up a short time ago in making a large cylinder. Says they have sent off about 2 Engines since he has been here & have 4 in hand. Employ their pattern makers to erect the Engines, Owen for one. John Hussey the smith who left us sometime ago, is not here & he knows nothing of him. John Hughes he says has had two letters from his father since he came, but is very close & communicates nothing of their Contents, except that he said that Dixon had called at our foundry.

I presume that by this time, Murdock has gained a harvest of intelligence. You will see that it will be proper for us to inspect the place whence the sand is got & send you a further dose of this nostrum. We shall also endeavour to learn something more of the secrets of this cupola & obtain an inspection of the [sand] & also that of the coal dust.

If anything has occurred, you may safely direct a letter to me here, as I shall have work for some days. Tomorrow Mr. Gott goes with us to inspect some situations in the neighbourhood.

I beg my duty to my father & kind respects to Yours & am truly
Yours &c   J. WATT JUNR.

Fenton is in London & not well. Murray at Newcastle. Halligan has applied to be taken back.

Holograph letter, J. Watt, Junr., to [Matthew] Robinson Boulton, Esq., Soho, near Birmingham.

Leeds 14 June 1802.

Dear Sir,—I have yours of the 11th. I mentioned in a postscript to my last that Halligan had applied to be engaged. The application was made to Murdock at the close of their drinking bout on Saturday, who advised him to come the next morning & speak to me. He came early yesterday (Sunday) and after stating his wife's uneasiness & ill health and glancing at his own disappointment, proposed to be readmitted into the pale of grace. This led to an explanation of his motives for leaving, to which he appears to have been instigated by Dixon, who wrote more than one letter to him. He said also that *others* had prompted him & that some disagreement he had had with one or two of his shopmates had made him indifferent about remaining at our Works. He seemed averse to entering into the details and I did not conceive it necessary to press him at that time. I told him that under the circumstances of his Wifes ill health, his former good conduct and my belief that he had not been instrumental in seducing others, I should be inclined to relax a little from our accustomed rigidity and use my Influence with you to take him back, but as he had shewn so little confidence in me before, in his manner of leaving us, I should expect him now to atone for that by leaving the terms of his return entirely to me. This he agreed to, & said he could not expect any other terms than we might please to allow him, or usually gave. I thought it necessary to proceed in this manner, as his wife had previously stated to Murdock that he was resolved to enter into no Articles. I appointed him to call again this Evening & I have got an Agreement ready cut & dried for him, to which I make no doubt of getting his signature. I propose to give him a guinea per week, as I think it politic that he should come back satisfied. I shall also lend him what money may be necessary to start him & pay Dixon three or four Guineas which he had borrowed, or rather which the other cunningly pressed him to borrow. It may then become a question of policy whether we should take him back or leave him for some time as a spy upon Murray & Dixon's proceedings.

I have been this morning with Halligan's wife who certainly looks miserably. She seems delighted with the prospect of returning and hates Dixon cordially. I hoped by her means to have got his letters, but she assures me in a way which makes me believe it, that they were carried about by her husband in his pocket until they were worn out & then committed to the flames. She says that young Hughes says he is informed by his father that in consequence of Dixon's call at our foundry they would soon see plenty of their old shopmates here. Hughes is very close & she could learn no more from him than this. She had never heard of the five guineas given him by Murray, but believed he had a guinea per week wages & was to rise at the end of every Year. That he was not yet engaged, but was to be so upon Murray's return. She showed us his bedroom, and as my key fortunately opened the trunk, we had a complete examination of its Contents, among which a roll of drawings of various parts of our Machinery & Engines deserves most conspicuous Mention. The things are very indifferently done, but the dimensions are written both upon the drawings, & upon separate slips of paper. There was about $4\frac{1}{2}$ Guineas, the remains of his fee probably, in his breeches pocket but no letters. Nancy says he always carries them about with him & and neither her nor my ingenuity can suggest any means of coming at them, other than that of seizing his person, which I had therefore resolved to do, but upon enquiry of Nicholson & Upton this morning, I find it can only be done by a Warrant from a Justice in the district in which the offence was committed. I therefore inclose the indenture, that *my father* may upon the strength of it procure a Warrant from Mr. Lane, which will be backed by one of the Justices here upon my swearing to the signature & the Gentleman will be apprehended, when we can at the same time seize upon his drawings & letters. It does not appear that we can make any legal use of the fact of his having made these drawings, as it does not appear that any of them are our property and perhaps he will be entitled to claim them back from us. To avoid this & also to prevent his disposing of them otherwise during the exchange of our letters, I was strongly inclined & indeed advised, to have laid hands upon them, but after having debated the subject, I am of opinion it will be better to leave them where they are as I do not think he will discover or suspect that they have been inspected & if they were removed it might lead to a premature discovery of our operations. They are not indeed of much consequence as they contain nothing but what Murray already knows, or soon may know through other channels. Hughes has no griefs to alledge against us & has been evidently instigated by his father, who intends coming also as soon as his time is out. He, however, speaks disrespectfully of us & the place, and I have no hope of bringing him back to his allegiance, at least whilst they continue to flatter & coax him here, which they naturally will do until his articles are signed. Under these circumstances there seems little advantage likely to result from putting the law in force against him, but I am not aware of any other mode of getting at his fathers letters, and it will at all events be some punishment to him to be sent handcuffed under the care of the Leeds constable & it will cost him some money to produce the proof of his being of age & in the mean time, he will probably be confined. It may also seem in some degree to criminate his father & enable us to commence those proceedings against him, which I am convinced we cannot postpone & which I hope we shall be able to do effectually from the additional evidence which Dixons recent embassy may furnish. It is certain that he went in search of three or four apprentices who have absconded from Murray, for which purpose he volunteered his services, having no doubt the double object in view of employing his talents in the seduction of our men. I do flatter myself that we shall be able to get compleat evidence against this scoundrel & his employers. I of course shall not see any of them, (indeed Wood is the only one within reach) unless the denouement of this affair should require it. I am in treaty for the Malthouse through the medium of a maltster & if it can be had, shall certainly make the purchase without waiting for farther communication, as I am pretty confident that the possession of it must enable us to dictate terms. I also expect an answer respecting two other plots of land immediately contiguous, which I think we ought to have, as they seem eligible speculations independent of the *ignoble* motives which dictate their purchase. Fenton's character is not very respectable. His brother is an Attorney well skilled in all the nefarious practices of his profession. Give me the Soho opinion about purchasing the land, the other I shall take for granted, unless I hear to the contrary before I have compleated my intentions, or the price proves too extravagant.

Fame has not outdone the magnitude of Murray's new Edifice, it is a rotundo of about 100 feet in diameter with a magnificent Entrance. The Engine is to stand in the middle and the lower rooms to serve as deposits for Engines & other finished goods. The Upper rooms to be for fitting. It is an excellent building & will not look amiss. It is up to the top of the 2nd storey. I will make a better sketch of it before I leave Leeds than this, but am now much pressed [Sketch, not reproduced] for time. He employs altogether about 160 Men the greatest part of whom are engaged. Many are Apprentices. He has much trouble to keep them & has been left by several.

I am in some haste   Dr Sir

Yours &c   J. WATT JUNR.

———

Letter from James Watt, Jnr., Leeds, to M. R. Boulton, dated June 15th, 1802.

Halligan has signed the agreement. . . . If I mistake not he has it in his power to benefit us most materially, as he has been extremely attentive to all that is going on in the foundry here and has picked up much valuable information. He is to remain with Murray as long as we may direct and to make application to try his hand at the green sand. He has promised to endeavour to get at old Hughes's letters upon Wednesday night when the youth goes to the play and it is supposed may leave his letters in his working clothes. I confess that this is not very probable from the caution he observes and if it does not succeed, must have them examined whilst he is drunk or sleeping, to ascertain whether they are worth taking. . . . The engines which Murray has now in hand are small ones 10 to 20 horse but a blowing or regulating cyl^r. has been made within a short time, for a person at Newcastle-upon-Tyne, whose name he thinks is Fishwick? His cutter block is pushed forward upon the boring rod by an endless screw, which, or some similar contrivance we must adopt, both to guard against the negligence of the borer and to save part of his wages.

Murray's dry sand foundry is about 20 yards long and 12 wide with two air furnaces and 3 stoves, one 20 feet by 13 wide for loam, another 17 feet by 13 for boxes and a third 17 feet by 9 for cores. The green sand foundry is on the opposite side of the yard about 15 yards distant & is nearly of the same dimensions, with two air furnaces and a cupola, but no stove, the cores being dried in those of the dry sand.

The cylinders above 20 horse are done in loam by the piece, by a very good moulder of the name of Joseph Brookes who has under him three men and a lad. The dry sand is done by Dixon with three men, of which Halligan is one, and a lad. In the green sand there are 4 men mounders and six lads. There are 5 chippers and 2 air furnace men and one cupula tender, in all for both foundries. The loam chip their own goods. The dry sand work is very badly done & much patched. The green sand & loam are both capital. There are 8 smiths with a similar number of strikers. Of the fitters he knows little, and has seldom been in their shops. He thinks that in both foundries they cast upon an average 3 to 4 ton per day, which I think is as much as we do (of this 50 cwt. is green sand). It is high time we should reform. The air furnace men have a guinea per week wages besides house rent free. The chippers have 18/- per week. There are 4 bookkeepers or clerks. The cash keeper and his assistant, who I suppose keeps the waste book & journal, a clerk who overlooks both foundries in the capacity of a A. Storey [Abraham Storey], a time keeper who also assists in paying the men & regulates the police of the yard. Fenton is generally busied in the counting house and probably keeps the ledger & writes the letters. Wood & Murray when at home, attend chiefly to the fitters. The men are paid upon the Saturday about 11 o'clock by the clerks who go into each shop, by which means their wives are enabled to go to market early. This regulation is much approved of by the men & is worthy of imitation for several reasons which I shall state when I see you."

Letter from James Watt, Jnr., Leeds, to M. R. Boulton, dated 17th June, 1802.

Halligan has inspected Hughes letters—nothing of importance.

———— There is little chance of getting more than one or two of Murray's men, as all the good hands are engaged. We shall have drawings and compleat information respecting his machinery. We have seen the engines erected by him here, which are certainly neat, but easily imitatable, in so far as they are worth imitation.

We also yesterday saw a manufactory of steam engines belonging to a Mr. Emmet at Birkingshaw nr. Bradford, where they bore their cylinders perpendicular for which they are about to take out a patent. They were putting up a blowing engine for their own furnace which was very well finished à la Murray. I saw the process of moulding the green sand, which was exactly that Halligan describes. One of our great defects is ramming the boxes too hard which prevents the escape of the air and steam. They scarcely ram theirs at all and make few or no prick holes in them. P.S. I propose returning by Halifax and Manchester, I have seen the place where the greensand is got which is exactly as described by Lawson. I wish you would write to Mr. Smith of Castleford (T. Wilsons uncle) to order 40 to 50 ton or say a barge load, which are equal to 2 boats.

We have also inspected the situation of an old furnace called Sea-croft about 4 miles from hence upon the road to Tadcaster, but do not think it offers any advantages.

I believe their (Murray's) coak comes from Bradford and has the reputation of being excellent.

Letter from James Watt, Jnr., to M. R. Boulton, dated 19th June, 1802.

. . . . I therefore propose to leave Upton (a solicitor) a Commission to purchase 1½ acre of the ground immediately contiguous to Murray and fronting the road at such price as he can agree. . . .

From what I have said in former letters you will gather that there are no hopes of getting at Murray's men, as all of the least consequence are engaged and at high wages. The only effectual way to harass them seems to be by destroying the basis of their illgot fame and setting up a competition which will diminish their orders. All our friends are of opinion that there is sufficient opening here for another works, and it does not appear that any of the present ones would be eligible connections : they are men without character & without means. The woman who has the malthouse will not sell at any price, she has set her mind on keeping it.

Letter from James Watt, Junr., to M. R. Boulton, dated 23rd June, 1802.

. . . . We saw at Rochdale an engine of Bateman & Sherrats newly erected & working very well & one of Murrays which was just set to work with the tappet wheels & usual finery. The cylr was 20 inches dr & they call it a 15 horse power. It burns much coal, which is a general complaint of every one Murrays engines & cost altogether about £1100. The owners are dissatisfied with it, as indeed they seem to be in every instance where we have gained access. It will be necessary to give the new finish to the ironwork of Mr. Taylor's engine, though he seems fully aware that the saving of coal is the essential object.

Before leaving Leeds instructed Upton to purchase only 1½ acres of front land if it can be had without the other—price not much to exceed £1000 per acre. Plot near the canal—price too high.

The green sand moulder (Davey) whom Murdock said *incog* was willing to travel south, but would not engage & would not work for any rate under 27/- per week.

Thinks Halligan will be a good instructor. Halligan is to stay on at Murrays unless he is found out.

Letter from James Watt, Jnr., London, to M. R. Boulton
21 Nov., 1802.

. . . I have had one conference with Weston respecting Murrays patent & he is to set to work upon the case as soon as we can be supplied with the following drawings—

I   a drawing of the nozzle at the Mint engine.

2   do   do   Mr. Symonds engine

3   do   old nozzles

4   do   eccentric circle & its application to the engine at Soho Foundry.

5   do   of other applications of the principle of opening the valves by the circular motion derived from the rotative shaft which have been devised or executed by us.

6   a complete drawing of engine with old nozzles & working gear & one with new manner.

When these are all ready it is proposed to lay the case before Holroyd & to give general retainers to Erskine and Garrow.

Has Lawson obtained any information from Garvin respecting the seduction employed by Murray & also respecting the conversation which the latter held in his presence & that of Mr Harason about the hollow spindle & his means of obtaining intelligence of our proceedings.

———

From letter of James Watt, Junr. [London] to Gregory Watt.
Dec. 7th, 1802.

. . . . 2 or 3 months ago we gave Mr. Swann an estimate for a 40 H.P. engine . . . . is told that the engine has been ordered from Murray, Fenton & Wood—whose estimate was £400 less.

———

Letter from James Watt, Junr., Soho, to M. R. Boulton, London, 6th June, 1803.

——— Weston I find is now resolved to take up the *Scire Facias* with proper energy. I have desired Murdock to put the finishing hand to the model which shall be sent you.

NOTE.—*Scire Facias* is a legal term meaning : " A judicial writ, for the purpose generally, of calling a man to show cause to the Court whence it issues, why execution of judgment passed should not be made out. This writ issues where execution has been delayed for a year and a day after judgment given."

———

From letter of James Watt, Junr., to Gregory Watt, 13 June, 1803.

Murdock's Model of his own & Murray's inventions is finished and shall be despatched by the first waggon.

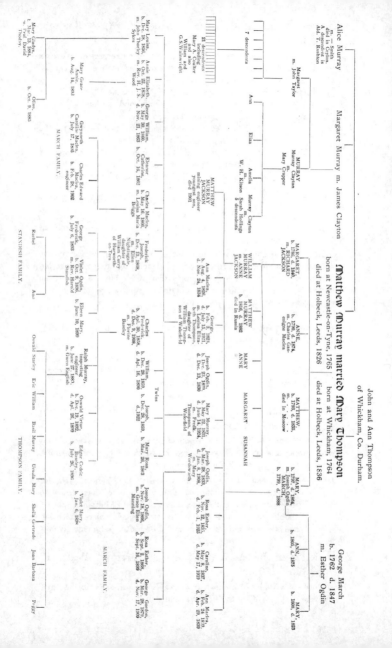

Matthew Murray married Mary Thompson

PART II.

Consisting of
Notes Chronologies, etc.,
by the Editor
and Contributions
by other Writers.

PART II.

THE WORKS OF FENTON, MURRAY & WOOD.

Extract from article in Rees' Encyclopædia, Vol. 34 :—

" Messrs. Fenton, Murray and Wood, of Leeds, Yorkshire, are the manufacturers of the most established reputation after Messrs. Watt and Boulton. The engines they send out cannot be excelled in beauty and perfection of workmanship, and they perform as well as any others. Their factory at Leeds is very extensive, and provided with every convenience for making all the parts of the engine in the best manner, and with the least labour.

They have three steam engines in the works, one for boring cylinders and turning large lathes, a second for turning small lathes, grinding, drilling the centres of wheels, tapping screws, &c., and for blowing the furnaces of the foundry ; and a third engine for working a great forge hammer, by which the heavy wrought iron work is forged.

The boring machines for cylinders, of which they have three in number, are very capital, as by an ingenious movement invented by Mr. Murray for drawing the bore (boring head?) through the cylinder, it is made to advance regularly, from one end to the other without any interruption. These machines are worked by a separate steam engine which is never stopped during the operation of boring a cylinder through, as it is found to make a sensible mark or ring if the motion is stopped. The best means are also taken to prevent the cylinder from changing its figure by its weight or by the pressure of the parts which hold it in its position.

The whole of the factory is lighted by gas lights in winter time. The boilers are manufactured by the aid of several machines to cut out the plate, pierce the holes, and bend the joints.

Before any of the smaller engines are sent away, all the parts are put together in a building on purpose where there are boilers fixed, and they are actually tried, to insure that every part is perfect ; they are then taken to pieces, with marks and directions for putting them together, and packed up for carriage, which is very easy, as there is a canal at the gates, which has communication by water to every part of England.

The above was written about 1816 by John Farey, a very able engineer of the period, who had often been in the works. (*See* page 52). In 1825-6 he was at Marshall's Mill, Holbeck, and was afterwards a consulting engineer in London.

In a letter in *The Mechanical World,* of December 3rd, **1887,** an old-time engineer who signs himself " H.," said :—

" I began my apprenticeship at Fenton, Murray & Jackson's, of Leeds, in 1832, who were at that time first class engineers and making engines of a superior quality to Stephenson's and not inferior to Boulton and Watts. I well remember there being 12 cranes in the works, some of them capable of lifting 10 tons, and a hydraulic crane capable of lifting 40 tons, for loading boilers."

NOTES BY J. OGDIN MARCH.

The following *Notes,* contributed to *Leeds Worthies,* published in 1865, are by Joseph Ogdin March, machine-maker, who married Mary, the third daughter of Matthew Murray. He was Mayor of Leeds in 1862.

He went to Scotland Mill, near Meanwood, to work for Mr. Marshall, in 1789, and at that age (twenty-four) made several valuable improvements in flax-machinery ; for which he was rewarded by a present of £20. But for his improvements at that time, it is nearly certain that flax-spinning in this neighbourhood would have ceased to exist, as all those embarked in it had lost the greater part of their capital without any success. Its establishment in Leeds was mainly due to his skill and ingenuity.

In his patent of 1799, in order to save fuel, Mr. Murray proposed to place a small cylinder with a piston on the top of the boiler, connected by a chain to the damper on the chimney, by means of which the force of the steam within the boiler had the effect of increasing or decreasing the draught of the fire, so as to keep up a regular degree of elastic force in the steam.

Mr. Murray also thought some advantage would be gained by placing the steam-cylinder in a horizontal instead of a vertical position, with a view of rendering the engine more compact than the usual construction ; he also adopted a new method of connecting the reciprocating motion of the piston-rod to a rotatory one of equal power, by means of a property of the rolling-circle, and showed how to fix the wheels for producing motion alternately in perpendicular and horizontal directions.

The slide-valve was first applied to the steam-engine by Mr. Murray in 1799, which answered the purpose of opening and closing four steam-passages, to use Dr. Robinson's words, " in a beautiful and simple manner," and he may be fairly considered the inventor.

He invented a very ingenious mechanism for adjusting the supply of air to the boiler-furnace, so as to diminish the quantity of smoke. This self-acting apparatus is described in the *London Journal* for 1821.

Fenton and Murray were the manufacturers of the most established reputation after Messrs. Boulton and Watt ; the engines they sent out could not be excelled in beauty and perfection of workmanship. Their extensive manufactory was provided with every convenience for making all the parts of the engine in the best manner, and with the least labour. They had the reputation of employing a greater quantity of tools, and of better and more ingenious construction, than any house in the trade.

He constructed a large amount of engine and mill work for the Russian Government, and had the honour of receiving from the Emperor a most valuable diamond ring. For works done for Sweden, he had also presented to him a gold snuff-box by the King of Sweden.

At his commencement, mill-gearing was in a very rude state ; he left it in nearly the state it is at present. The large framings for the first motions of mills are to this day models of elegance, possessing everything requisite for strength and durability. He touched nothing that did not come out of his hands a new thing. Considering that he was no mathematician, it was truly surprising that the sketches of drawings which he wanted making were remarkably proportionate; showing the strengths very nearly accurate when they were reckoned out.

He had the attribute of real genius—a truly liberal mind! nothing pleased him more than to exhibit the great stores of his rich mechanical mind to a kindred spirit. For clever tools and implements, and especially for the forgings of beat-iron work, such as parallel motions and the like, he was far in advance of others.

A memorable instance of his liberality was shown by the invitation he gave to Mr. Murdock, the managing partner of Mr. Watt, to stop a week at Steam Hall,—Mr. Murray having built a very handsome house, which was called Steam Hall, because it was heated entirely by steam. Mr. Murdock accepted the invitation, and had free access to every part of the works, and every attention was shown to him.

For this kindness Mr. Murray received a most ungenerous return, for, immediately afterwards Messrs. Boulton and Watt bought a large field near his works, for the express purpose of preventing their extension. Nor was this the only source of complaint. On his way from London Mr. Murray called at Birmingham for the purpose of looking over the Soho Works, and enjoying a treat in examining the tools of Mr. Watt's establishment. He was received politely, together with Mrs. Murray, who accompanied him, and both were invited to dine; Mr. Murdock, however, hoped they would excuse him declining to show him the works, as their rule was not to show them to any persons in the trade. It need hardly be added that Mr. Murray was greatly affronted, and at once declined the offered dinner.

Mr. Murray lighted this town with gas at a very early period in the history of gas manufacture; introducing many improvements in the construction of the retorts, the condensers, and the various other parts of the apparatus.

J. Bentham, R. Roberts and F. Fox have been credited with the early metal planers, but it is probable than Fenton, Murray & Wood had the first, in order to make flat surfaces on the D slide valves which Murray used. J. O. March informed Dr. S. Smiles that when he first went to work for Murray in 1814 he saw the planing machine.

" I recollect it very distinctly, and even the sort of framing on which it stood. The machine was not patented; and like many inventions in those days, it was kept as much secret as possible, being locked up in a small room by itself to which the ordinary workmen could not obtain access. The year in which I remember it being in use was, so far as I am aware, long before any planing-machine of a similar kind had been invented."

No. 1752, of 1st June, 1790. Machine for spinning yarn from silk, cotton, hemp, tow, flax and wool.

NOTE.—This invention helped to revolutionise the flax spinning Industry.

No. 1971, of 18th Dec., 1793. Instrument and machine for preparing and spinning flax, hemp, tow, wool and silk.

NOTE.—This specification includes an improved Carding engine.

No. 2327, of 16th July, 1799. Steam engine.

NOTE.—This shows an automatic method of regulating the draught and the steam pressure of a boiler.

No. 2531, of 11th Aug., 1801. Constructing the air pump and other parts belonging to steam engines so as to increase power and save fuel.

NOTE.—A form of mechanical stoker is shown.

No. 2632, of 28th June, 1802. Combined steam engines for producing circular powers, and machinery belonging thereto, applicable for drawing coals and other minerals from mines, for spinning cotton, flax and wool, or for any purpose requiring circular power.

NOTE.—This shows the box or three-ported slide valve.

No. 3792, of 12th March, 1814. Constuction of hydraulic presses for pressing cloth and paper and for other purposes.

NOTE.—The hydraulic ram and the top table move in unison, and a pressure indicator is shown.

Patent 2531 of 1801, was set aside by *Scire facias* at the instance of Boulton & Watt, on the ground that a certain portion was known to them. It shows a circular motion for operating steam drop valves by wipers or cams, and was the first time such a device had been illustrated in a patent specification. Benjamin Hick & Sons, of Bolton, afterwards adopted the method for large steam engines.

There was a claim for drop valves which Murdock said he had used. Murray lost the whole Patent, although it was acknowledged to contain much that was entirely novel. For example, one claim was for a device for preventing smoke, and this came into use many years afterwards as a "mechanical stoker." In those days Patents cost a great deal of money in fees, etc., and it was usual to cover a number of things in one specification.

Under the present Law a Patent can no longer be unfairly invalidated in case it is found to contain something done before. That particular part or claim is merely deleted or modified and the rest stands valid.

FLAX HACKLING MACHINE.

In 1809 Matthew Murray sent a model of a machine for hackling flax to the Society of Arts, and for it he was awarded a Gold Medal, which was presented to him personally by the Duke of Sussex, afterwards President of the Society.

The Transactions of the Society for 1809, Vol. XXVII., pages 148-150, announce the award as follows :—

> "The Gold Medal of the Society was this session voted to Matthew Murray, Esq., of Leeds, for a machine for Hackling Hemp or Flax. The following communication was received from him, an explanatory engraving is annexed, and a complete working model is preserved in the Society's Repository."

The first letter written by Murray when submitting his invention is dated February 10th, 1809, and the second, written two months later, gives the following description of his machine :—

> "The flax is brought perpendicularly down upon the teeth of the hackles. The hackle moves through the flax in the direction of the fibres, and is only suffered to penetrate as the fibres become open. It reduces the operation to a certainty; which was not so before, which is of material importance for mill spinning, as flax does not undergo any other separating process after being hackled."

The model is now owned by Mr. Ralph Murray Thompson, who has loaned it to the Science Museum, South Kensington.

MODEL OF FLAX HACKLING MACHINE.

BORING MACHINES.

In a paper on the " Early History of the Cylinder Boring Machine," by E. A. Forward, M.I.Mech.E., in the transactions by the Newcomen Society of 1926, he gives the following information about Murray's machines :—

" After the date of Billingsley's patent, the author has found no contemporary illustration of a cylinder boring machine until 1817, when John Farey made a drawing of one of Matthew Murray's machines for the sixth edition of the ' Encyclopædia Britannica.'

The machine has a solid bar with two longitudinal grooves along opposite sides of it, and in each of these grooves there is laid a rack bar with its teeth facing inwards. The racks are fixed to the sliding cutter head, and revolve with the bar, while the rack teeth, which virtually form a long nut, engage, beyond the end of the boring bar, with a short screw mounted on a central spindle.

This spindle extends rearwards so as to allow the racks to attain the full travel of the head, and its outer end is connected by gearing and a long upper shaft with a spur wheel mounted on the end of the boring bar. The bar is 8.5in. diameter and 9ft. long, and is shown boring a cylinder 17in. diameter by 5ft. long.

As Murray started his engine factory at Leeds in 1795, this machine may have been constructed then, or shortly afterwards. Farey, writing in 1816, said that Murray had three machines for boring cylinders. He is also known to have been making boring machines for export as early as 1806.

Wilkinson's rack feed machine and both Murray's and Dixon's machines were practically twice as long as their bars. Murray's, however, was superior in that it had a positive automatic feed, while Dixon's had only a hand feed.

The next machine to be noted is of another design, also attributed, by later writers, to Matthew Murray. In this the cutter head is fixed to the middle of a very long boring bar, which rotates and at the same time slides through its two bearings by the agency of a screw and nut. Although brief descriptions of this form of machine are to be found in English works, the earliest complete drawing of one, extracted from a French publication, was given in Gill's *Technical Repository* for 1822.

The machine described was in ise at the engine works at Chaillot belonging to M. Périer, and was said to be the machine which the English mechanics then used for the purpose of boring cylinders. It is stated that only two such machines existed in France, the one at Chaillot and another at St. Quentin. It is highly probable that both these machines had been made by Murray, who was one of the principal machine tool makers of the period.

The French writer evidently appreciated the positive automatic feed, as he stated that ' the movement of tl ѳ boring head being uniformly continued, the result is that the work is more quickly performed, and is more perfect than when done on the old machines where one was obliged to raise, from time to time, the weight which impelled the boring head.'

A much better drawing of the Chaillot machine was published, however, in the ' *Bulletin de la Société d'Encouragement,*' in 1823. The machine had a boring bar 5.5in. diameter and 13.25ft. long. Its two bearings were 6ft. 4in. apart, and the maximum travel of the head was 5ft. The boring head was keyed to the middle of the bar and the whole bar was moved endwise through its two bearings and through the driving wheel, which was mounted close up to one of the bearings and the key of which fitted a slot cut along the bar.

A screw, 1.75in. diameter and 6ft. long, with a pitch of 0.5in., was fixed to the end of the bar and had its outer end supported by a nut which could turn in a fixed bearing. The nut had a gear wheel with thirty-eight teeth fixed to it, which engaged with a 36-tooth wheel fixed to the end of a square countershaft, along which could slide a second wheel with thirty-eight teeth that constantly engaged with a wheel of thirty-six teeth fixed on the end of the boring bar. The sliding wheel was pushed along by a flange on one side of it.

The bar was driven from a vee-grooved stepped pulley through a double reduction gearing. As the bar rotated, the feed gearing caused the nut to rotate relatively backward at about one-eighth of the speed, so giving a feed of 0.065in. per revolution of the bar. The bar being drawn through the driving wheel, held the latter in position against the bearing. A cylinder 20.5in. diameter and 4ft. long is shown on the machine. It is supported on adjustable vee blocks, to which it is secured by chains with tightening screws.

Machines of this form were made for sale and export, and not merely by an engine maker for use in his own works. Simon Goodrich had a small machine of this type, no doubt made by Murray, in Portsmouth Dockyard in 1813, and he gives a rough sketch of it in his ' Journal.' "

The Figure on the next page shows Murray's boring machine as used in his works. It appeared in the 6th Edition of " Encyclopædia Britannica " of 1817.

An Essay on the Framing of Millwork, pub. by Robertson Buchanan, of Glasgow, in 1808, states that—

> Mr. Murray, of Leeds, has found a way of boring the hardest cast iron. He shewed me some cylinders of it, which from the whiteness of the grain at the places broken off by a chesil, denoted its quality. Such iron is equally hard throughout.

> He mentioned that a patent was granted for wheel bushes of metal, of a peculiar hardness, which metal is nothing more than very hard cast-iron, but the Patentee had discovered a mode of boring it.

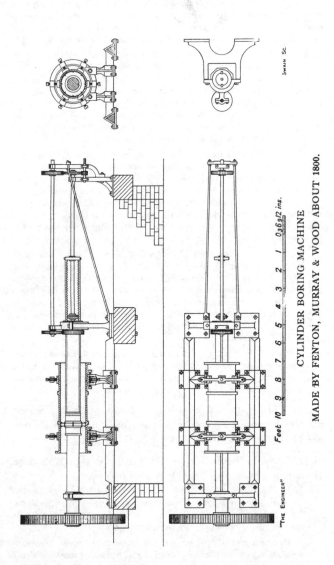

Swain Sc.

Feet 10 9 8 7 6 5 4 3 2 1 0 36 9½ ins.

"The Engineer"

CYLINDER BORING MACHINE

MADE BY FENTON, MURRAY & WOOD ABOUT 1800.

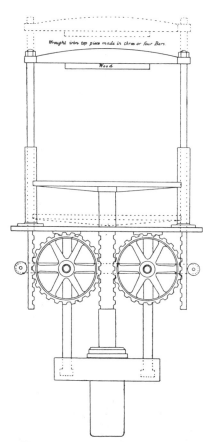

Wrought iron top piece made in three or four Bars

Wood

Murray's Patent 3792 of 1814, contains a drawing of a machine for pressing woollen goods, etc., in which the top and bottom tables approach each other at the same time. This admitted of considerable extension between the top and bottom without raising the bale, when pressed, too far from the floor.

A press like this was supplied to Gott's mill at Bean Ing.

Chain Cable Testing Machine.

The Goodrich collection at the Science Museum has an original drawing of a Chain cable proving machine made by Murray in 1826. His writing is on the drawing.

The ram is one foot diameter and the cylinder, which is horizontal, measures 2 feet diameter and $4\frac{1}{2}$ feet long. The total length of the machine is 34 feet.

Two crossheads rest on rollers which run on two tables and above the cylinder there is a compound lever gauge to measure the water pressure, also at the extreme end a long lever to measure the direct load. This was a forerunner of many hundreds of hydraulic testing machines made by Greenwood & Batley Ltd., Joshua Buckton & Co. Ltd., and others.

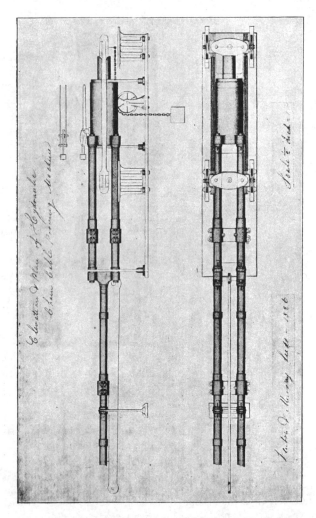

HYDRAULIC TESTING MACHINE TO PROVE CHAIN CABLES
UP TO 1,000 TONS, FOR GENERAL WILSON.

NOTE.—Copy of the drawing in the Goodrich Collection at the Science
Museum, South Kensington, referred to in letters which
Matthew Murray wrote to the Engineer and Mechanist of the
Navy Board on 16th April and 21st November, 1824.

ABSTRACT FROM MATTHEW MURRAY'S PATENT No. 3742
OF 1814; FOR AN INDICATOR TO ATTACH TO THE
EXTERNAL CYLINDER OF A WATER PRESS.

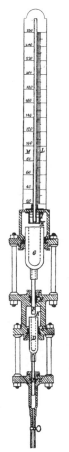

A, copper pipe leading to the small cylinder B, against which the pressure of the water acts in proportion as its area is contained in the area of the internal cylinder C; D, a cylinder attached to the cylinder B, and moving freely up and down with it in the leather E, E; F, a small cylinder, one-tenth of the area of the cylinder D, and firmly attached to the last or largest cylinder G, and moving up and down freely with it in the leathers H, H; I is a small iron cistern containing mercury, fixed to the inside of the receiver K, with holes near the top to admit the pressure of the water contained in K, to act upon the mercury in I; L, M, a glass tube and graduated scale rising by tons from twenty to two hundred and sixty, but any other proportion might be used where smaller or larger pressures are necessary. In this case the proportions of C to B are as four hundred to one; the proportions from D to F are ten to one; the diameter of G is four inches, and when forcing the water into the receiver K, will have to act against a resistance equal to the weight of a column of mercury of its own diameter; the height of the mercury showing the weight in tons in the scale M for small pressures; and for weighing heavy goods I omit the small cylinders D, B, and make use only of the cylinder motions G, F; it is necessary to observe that these cylinders must be perfectly smooth and true, made of good bell metal, and move up and down in the leathers with the least possible friction. D and G are also hollow, to admit the small piston F and iron cistern I. The receivers K, K, must be kept full of water, but any small leakage in the leathers will not alter the power of the instrument for indicating the pressure on the materials acted upon by the press.

Indicator for
Measuring
High Hydraulic
Pressures.

MECHANICAL STOKER.

Patent No. 2531 of 1801 is of special interest because the specification describes a method of making a fire-grate for consuming smoke. This patent was opposed by Boulton & Watt, but not on account of the patent fire-grate.

The drawing shown below has a cog wheel and a rack for mechanically pushing the coal forward, and the device is thus a form of *mechanical* stoker. It was the first of its kind.

Mr. David Brownlie, B.Sc. (Lond.), F.C.S., a recognised authority on the subject of mechanical stoking, says :—

"It is entirely new to me that Matthew Murray did this work, and I shall most certainly mention it in my next publication. The more one learns about the early history of engineering the more amazing does the work of these early pioneers appear."

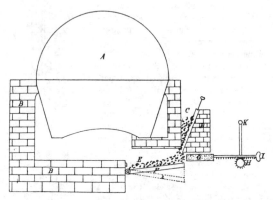

A, the boiler; B, B, B, B, brickwork; C, the tunnel for conveying coals to the fire; D, a sliding plate for lessening the aperture of the funnel, and regulating the quantity of coals in going in; E, the fire-place; F, the grate barrs, lying in a frame, and running upon a joint, the grate barrs lowered down for the purpose of cleaning out the fire; G, charging box for pushing the fire farther on the grate barrs, and making room for fresh coals as the other is consumed; this box is filled with holes, for the purpose of letting air into the fire; H, a small wheel and rack for working the charging box G in large fires, but the box may be thrust in by the handle I, in small ones; K, a lever for turning the wheel H.

In 1825 Murray designed an articulated locomotive which included a mechanical stoker. *See* page 65.

STEAM BOAT ENGINES.

It has been usual to give Fulton, of America, the credit for commercialising steam propelled vessels, but John Wright and Matthew Murray are also worthy of praise, because of their pioneer work. In 1813, a French privateer called L'Actif, which had been purchased from the Government by John Wright, a Quaker of Yarmouth, was brought to the Canal Basin, Leeds, and an engine and boiler fitted into her by Murray's firm.

The boat was 52 ft. long, clinker built for three long sails and twenty oars. The engine that Murray fitted had a cylinder 8 inches diameter and 2½ feet long, and it gave 8 horse power when supplied with what was then called " high pressure " steam—probably about 30 lbs. per sq. inch.

The boat was fitted with paddle wheels and re-named the " Experiment," and had her first public trial on the River Aire at Leeds Bridge. She went under steam to Yarmouth, and on August 9th, 1813, made her first public trip, carrying passengers from Turner's Bowling Green to Braydon, as reported by the Yarmouth correspondent of the *Norwich Mercury,* as follows :—

> " On Monday (August 9th) the first experiment was tried with the Steam Packet Boat, on which occasion Sir Edmund and Lady Lacon and family with a party of ladies went in the boat unto Braydon, and expressed themselves highly gratified with their excursion. She afterwards went through bridge amidst the acclamation of thousands of spectators. The boat has since gone regularly to and from Norwich and answers every expectation, and we have no doubt but that she will reward the exertions of her spirited proprietor and projector."

Eventually Murray's engine and boiler were removed from the " Experiment " to " The Courier," which after running alternately with another boat called " The Telegraph " for about nine months, was steamed to the Medway, and for seven weeks plied between Sheerness and Chatham. It was the first steam boat to make a voyage *to sea* in this country.

Murray also built two large engines for Francis B. Ogden, who was for several years United States Consul in Liverpool. He is famous as being the first man to establish a paddle steam boat service on the Mississippi River, and he did this with a boat fitted with one of Murray's engines made in Leeds in 1816. See *Mechanics' Magazine,* Oct. 30, 1830.

In 1813, Ogden obtained a patent in the United States, which claimed for what Murray had done two years previously.

> " The application of the powers of the pistons in such manner as they shall act simultaneously on the same axes at right angles with each other. When one is at its minimum the other being at its maximum will carry it past the point, and that they will mutually tend to equalise the motion, thereby doing away the necessity of balance wheels, etc."

The engines had two cylinders each, 33 inches diameter and 4 feet stroke, and each cylinder actuated a separate beam, the other ends of the beams being connected to cranks at right angles. The beams were supported on A frames of the special design standardised by Murray, and this design was continued on many paddle boat engines built in the U.S.A.

The second engine was placed on a Mississippi steam boat,, the forerunner of those described many years later by Mark Twain in " Huckleberry Finn " and other stories. The Editor of *Mechanics' Magazine* quotes Mr. Ogden :—

> " He had frequently seen the boat ascending that stream against a current of three miles and a half with a ship of three or four hundred tons on each side, and smaller vessels, tugs and schooners, astern."

On 17th July, 1815, Murray wrote to Simon Goodrich :—

> " They might be used with great advantage in His Majesty's Navy if properly introduced . . . two 20 or 30 Horse cylinders be applyed on Board a Gun Brigg . . . As these will be adopted sooner or later in the British Navy."

The side lever type of engine which Murray was the first to introduce and which Francis Fox, of Derby, copied, was later adopted extensively for marine purposes. (*See* page 62).

James Watt did not approve of propelling vessels by steam engines, and in 1800 wrote to Henry Bell that it was " pure waste of time."

At the end of the biography of James Watt in the " Dictionary of National Biography," it is stated that a steam boat owned by James Watt, Junior, was the first to make a sea voyage from a port in this country. The above-mentioned " Telegraph," with a Murray engine and boiler, had already been to sea from an English port !

MODELS IN THE SCIENCE MUSEUM.

The Authorities of the Science Museum, South Kensington, consider Matthew Murray's inventions of sufficient importance to have had models made of them. The following notes are from the Official Catalogue and by Mr. H. W. Dickinson, M.I.Mech.E., Curator and Honorary Secretary of the Newcomen Society :—

DRAWING OF FENTON AND MURRAY'S ENGINE. (Scale 1 : 8.) Prepared in the Museum, 1897.

Shows the self-contained or " portable " engine patented in 1802 by Matthew Murray for producing " circular power." Several engines of this type were constructed by Messrs. Fenton, Murray & Wood, of Leeds ; the one represented was of 4 h.p., and was erected in 1802 for grinding bark at a Bermondsey tannery. The arrangement is exceptionally compact, owing to the shortness of the virtual connecting rod employed, its length being equal to one-half of the stroke only.

The engine has a vertical double-acting cylinder, with an air-pump and jet condenser in a tank below. The valves are of the drop type, and are driven by tappets on a revolving shaft connected with the crank-shaft by bevel wheels.

MODEL OF MURRAY'S ENGINE (WORKING). (Scale 1 : 10.) Made in the Museum, 1902.

Shows the mechanism adopted in the self-contained engine, patented in 1802 by Matthew Murray, for converting reciprocating motion of the piston into rotary motion of the shaft. It is an application of the fact, originally published in 1666 by De la Hire, that when a circle rolls within another of double its diameter, any point in the circumference of the rolling circle describes a straight line. Murray appears to have been the first to construct practically a machine embodying this method of obtaining a straight line, or " parallel " motion ; he, moreover, fully describes a means of taking up the wear on the centre of the rolling wheel by the use of adjustable double cones.

On the fly-wheel shaft is a crank pin having loose on it, a spur wheel of equal radius to the crank path, while on the rim of the spur wheel is secured a crank pin to which the piston-rod is directly attached ; the spur wheel engages with an internally geared rim of double its diameter rigidly secured to the engine framing. By these arrangements a guided straight line motion is obtained without the use of links or guides, while, owing to there being no extension of the piston-rod or anything beyond it, the engine is remarkably compact.

Murray's mechanism possesses the further advantages of giving a true harmonic motion, and having all its parts revolving so that they can be directly balanced, while there is a total absence of oblique thrust. For these reasons it is, although now no longer used in steam engines, still extensively adopted for giving reciprocating motion to the tables of flat-bed printing machines as shown by a model in another Section.

BEAM ENGINE. Received 1907.

This form of " portable " or self-contained condensing beam engine was brought out in 1805 by Matthew Murray, of Leeds. It proved in practice to be not easily accessible, and only a few engines were therefore made. It is interesting, nevertheless, as having the same general arrangement as the side-lever engine which subsequently became the recognised type for marine purposes.

The engine is contained within a tank supported on four feet. A dividing partition enables the cylinder and valve box to be packed with non-conducting material, while the air-pump and condenser are surrounded by water. The cylinder crosshead has two side rods working on to a pin on the end of a beam rocking on a centre on the underside of the tank. The other end of the beam has a connecting rod to a crankshaft above, with a large fly-wheel. The air-pump is worked similarly by side rods from an intermediate point in the beam. Neither cylinder nor air-pump crossheads are guided.

Steam is distributed to the cylinder by the " locomotive " slide-valve which had been patended by Murray in 1802. The valve has a rack on the back geared into by a sector which is actuated by a rod having a frame at the other end embracing a cam on the crankshaft; the cam gives a period of rest at the end of each stroke. The injection cock to the condenser can be adjusted by a hand lever moving over an index plate.

The engines actually made were of 6 h.p. ; the small example shown has a cylinder 3 in. diam. by 5 in. stroke.

DESCRIPTION OF MURRAY'S LOCOMOTIVE.

The following description is by Mr. E. A. Forward, A.R.C.Sc., M.I.Mech.E., of the Science Museum, South Kensington, and it follows details of the model made by the Museum workshop in 1910 to one-eighth scale. This is the best model that has been made, and it was copied from a line drawing in " Bulletin de la Societé d'Encouragement pour l'Industrie Nationale," Paris, published in 1815. The model is a faithful copy in every detail, including feed water tank, gauge taps, and lagging on boiler.

The arrangement of the engine was an improvement on those built by Trevithick, although obviously based on them, in having, at the suggestion of Matthew Murray, two cylinders working on separate shafts so connected that the cranks remained at right angles, thus avoiding any difficulty in starting. There is reason to believe, however, from the contemporary records upon which the model is based, that the design did not arrive at once at its final shape.

The engine had a cast-iron boiler of oval section 37 in. by 32 in. and 9.6 ft. long, made in two parts bolted together, and having, a single furnace flue 14 in. diam. passing through it; in the boiler on the centre line were sunk, for one half their length, two vertical cylinders, 9 in. diam. by 22 in. stroke, exhausting directly into the atmosphere.

Each piston-rod was controlled by two vertical guides, while by a pair of return connecting rods it drove parallel outside cranks on a crankshaft below it. These two crankshafts were connected through gearing with an intermediate shaft, upon one end of which was a large spur wheel gearing with the teeth of the rack rails.

The steam distributing valves were four-way plug cocks oscillated through 90 deg. by wrist plates, which were connected by horizontal rods above the boiler, with vertical levers pivoted near the centres of the boiler ends; these levers extended an equal distance downwards, and their lower ends were connected with eccentrics mounted on the crankshafts.

Reversing was effected by making the cocks oscillate through an angle of 90 deg. adjacent to the angle used for forward motion, and this was done by attaching the valve rods to points in the wrist plates at right angles to the former ones. Short levers, having the valve rods attached to their lower ends, were mounted loosely on the valve stems, and pins in their upper ends engaged with either of two holes in the wrist plates. Forked hand-levers, engaging with collars on the valve lever bosses, were provided for sliding them into and out of gear. Two smaller plug cocks, coupled by a rod, controlled the steam supply from the boiler.

A direct loaded spring safety valve was fitted at each end of the boiler shell. The boiler and gearing were supported by a wooden frame carried upon four wheels 35 in. diam., with a wheel base of 7.33 ft.; the driving spur wheel was 38.2 in. pitch diam., and revolved at one half the speed of the crankshafts, so that the tractive factor was 93.6. The boiler was fed by a pump, immersed in a water tank carried at the front end of the engine, and driven by the valve gear. The fuel was carried on a platform between the frame beams, at the rear of which the driver stood, so that no tender was required. The boiler and cylinders were lagged entirely with wood.

Blenkinsop stated that one of these engines weighed 5 tons, and cost £400, and that it did the work of 16 horses in 12 hours. It drew 27 wagons, representing a load of 94 tons, at 3.5 miles an hour on the level, or 15 tons up a gradient of 1 in 18; lightly loaded its speed was 10 miles an hour. The consumption of coal was 21.3 lb., and of water 14.3 gal. per train mile, so that each pound of coal evaporated 6.7 lb. of water.

Although the official opening of the railway from Hunslet to Middleton was 12th of August, 1812, the first locomotive must have been made about a year before.

This illustration has on it a representation of the locomotive and the date 1811. Also Mr. Edward Lawson has a plaque dated 1811, which has pictures of Middleton and the locomotive.

The first locomotive of George Stephenson was built at Killingworth Colliery, in 1814, and it followed the design of Murray's very closely, the principal difference being that the boiler was of wrought iron instead of cast; and there were two sets of gear wheels between the two vertical cylinders and the driving wheels, whereas Murray's engine had one set to the cogwheel.

When the " Rocket " locomotive was made in 1828, it embodied improvements then know, including suggestions of Trevithick to Robert Stephenson.

Murray's locomotive was a complete working unit with spring safety valve, feed pump, water tank, etc.

	Murray's locomotive. 1811.	Stephenson's locomotive. 1814.
Diameter of piston,	9 inches.	8 inches.
Stroke,	22 inches.	24 inches.
Section of boiler,	Oval 32 ins. x 37 ins.	Round 34 ins. dia.
Length of boiler,	9½ feet.	8 feet.
Diameter of flue tube,	14 inches.	20 inches.

MURRAY'S LAST DESIGN OF A LOCOMOTIVE.

In October, 1825, Murray recommended to the Stockton and Darlington Railway a design involving the use of separate carriages for the engine and boiler, and this was published as a communication to *Newton's London Journal* in 1826, after Murray's death :—

> " Mr. Matthew Murray, of Leeds, whose great experience and superior talent in the construction of steam engine, the scientific world is well aware of, has proposed a design for a locomotive engine, which he considers will be found to possess superior advantages to any that have been heretofore employed. This plan for a locomotive engine, however, is to be considered but as a design, for though Mr. Murray believes it to possess advantages over all others heretofore made, or at present in use, yet he is far from presenting it as a piece of machinery that is not susceptible of improvement."

In the " Locomotive " of February, 1926, Mr. E. A. Forward says :—

> " In the design shown, the engine and boiler are carried on independent four-wheeled carriages, but all the wheels are driven. The wheels are of iron plate with flanged rims bolted on, and the frames are of cast iron mounted on springs. The two cylinders are placed vertically one on each side of the centre line near the middle of the carriage, and have long D valves, the exhaust steam passing to the chimney by a turned-up pipe. The piston rods project upwards and are connected with the ends of half-beams, the other ends of which are attached to the upper ends of vertical vibrating links mounted on the frame ends. The connecting rods are attached to points in the beams nearer to the pivots, thus reducing their stroke, and the two rods drive cranks fixed at right angles on an intermediate shaft below the frame. A toothed wheel on this shaft gears with an equal wheel on the front axle, thus driving the engine. The front and rear wheels of each carriage are coupled by side rods, while the rear axle of the engine carriage and the front axle of the boiler carriage are coupled by a chain and sprockets.

> The boiler is cylindrical, with a single flue having a water bridge and a water baffle across it. The flue turns upwards inside the shell and passes through the top plate. A damper is provided at the top of the chimney. The most remarkable feature of the boiler, however, is that it is fitted with a form of mechanical stoker ; a hopper above the grate is fitted with a feeding plate and roller which was to be driven by the running wheels."

Ten years afterwards the separate engine and boiler arrangement was patented by Mr. T. E. Harrison and adopted for the experimental locomotives, " Hurricane " and " Thunderer," for the Great Western Railway.

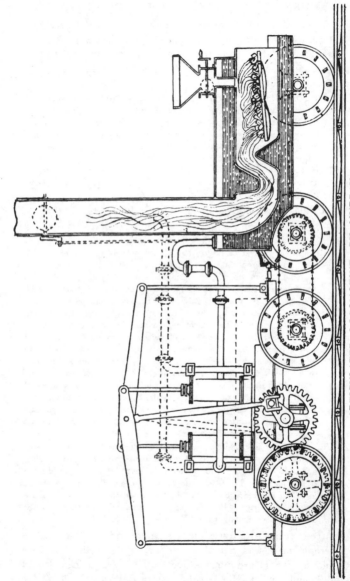

ARTICULATED LOCOMOTIVE, BY MURRAY.

He sent the design to Stockton & Darlington Railway Company in 1825.

66

The first commercially successful railway in the world was developed from a waggon-way laid down in 1758 from a coal staithe at Casson Close, near Leeds Bridge, to the "Day-hole" of a coal mine below the village of Middleton. It was authorised by an Act of Parliament obtained by Charles Brandling. The following abstract gives interesting particulars of it and the then price of coal, etc. :—

> WHEREAS Charles Brandling Esquire Lord of the Manor of Middleton in the County of York is owner and proprietor of divers coal works mines veins and seams of coal lying and being within the said Manor of Middleton and places adjacent and hath proposed and is willing to engage and undertake to furnish and supply the inhabitants of the Town of Leeds with Coals for their necessary use and consumption at the rate or price of 4¾d. a corfe containing in weight about 210 lbs. and in measure 7,680 cubical inches for the term of 80 years to commence from the second day of January 1758 and for such further term or longer time as the said mines or any of them shall continue to be used and wrought and at his own charge and expense to carry and convey or cause to be carried and conveyed from his said coal works yearly and every year 20,000 dozen or 240,000 corfes of coals at the least and to lay up and deposit such coals or cause the same to be laid up and deposited upon a certain field or open space called Casson Close near the Great Bridge at Leeds in order to be there sold and delivered at the rate and price aforesaid unto the inhabitants of the said town of Leeds or to such other person as shall purchase the same.

By a second Act of Parliament in 1779 the quantity of coal to be delivered to Leeds was raised to 48,000 tons per annum, and the price to 4/9 a ton; a third Act of 1793, raised the price to 5/9, and in 1803, 7/- a ton was the authorised price.

The waggon-way passed over a part of Hunslet Moor, across two public highways, and through the Leeds Pottery works; the rest of the way being across fields. The coal corves or waggons were hauled by horses and the line was unfenced.

In early days it was customary to make the wheel tracks of hard wood "stringers" about 6 inches square and 6 feet long, tied together by wood sleepers. Sometimes upper stringers were spiked to the lower, for renewal when worn.

About the end of the 18th century, cast-iron plates and rails began to be used, and these would be cast in the old Foundry at Carr Moor Side, Hunslet, close to the line. It was owned by Timothy Goddard and Titus Salt, and principally occupied with work for collieries at Middleton, Beeston, Rothwell, etc.

The proprietors of Leeds Pottery received a nominal rental of £7 a year for the line to run through the works, also a reduction in price of coal. It is on record that in 1800, they paid over £2,000 for coal; sold £30,000 worth of pottery, and the wages bill was £8,000.

The coal came from what is known as the Beeston top seam, which was mined in Beeston as early as 1570. It was brought out through a " Day-hole " situated at the top of a fairly steep incline down which the full corves were lowered by a ropeway which pulled up the empties.

The introduction of steam locomotion on the line in 1812, came about as a result of the Napoleonic Wars, raising the prices of all cereals. At one time wheat was over £8 a quarter. Horse feed became very expensive, and yet, to maintain the Charter the Colliery proprietor had to deliver a considerable quantity of coal to Leeds each year.

As is usually the case when there are abnormal conditions due to wars, etc., inventors and others became very active, and there resulted the great advances in utilising steam power, which characterised the earlier years of the last century.

One of the great inventors discovered in that period was Richard Trevithick who, to make steam locomotion possible, used steam at about 30 lbs. per square inch. It was called " high pressure " to distinguish it from pressures of only a few pounds then used in pumping and mill engines. These low pressure engines obtained their power by vacuum set up by condensing the steam.

Owing to the size and the weight of the low pressure cylinders for a given power and to the necessity of using condensing water in the large condensers, it was obviously necessary for steam locomotive engines to work on different lines. It was this requirement which helped to bring high pressure steam into general use.

Trevithick ran a single cylinder locomotive with flywheel on plate rails at Penydarran Ironworks, Merthyr Tydvil, in 1804, and in 1808 exhibited a locomotive (also having a single cylinder) on a circular track made of edge rails, where Euston Square now stands. The locomotive was seen by important people, including Charles Brandling, who being a Member of Parliament had to go to London regularly.

He would naturally talk over such an important innovation with John Blenkinsop, his colliery manager, because in the use of steam power they would see a way out of the difficulty of hauling by horses. As a result, Blenkinsop took out his patent No. 3431 of 1811 for a rack railway.

It must be remembered that it was then the " age of cast-iron," and as rails had to be made of that metal, the locomotive had also to be of light weight so as not to break them. Roughly speaking, a locomotive will haul on the level about four times its own weight by adhesion of plain wheels on smooth rails, and therefore one of 5 tons could only have hauled, say, 15 or 20 tons. Brandling and Blenkinsop wanted to haul much more on each trip.

By means of the rack rail, the 5 ton locomotive which was actually built did regularly haul 90 tons, and the man who made this possible was Matthew Murray, then the leading engineer in the North of England. Unlike James Watt, he was favourable to steam locomotion and to the use of higher steam pressures.

He must have had the order for the locomotives in 1810, because he had to make experiments, and it is known that he had at least one locomotive completed by the following year. There were representations of this locomotive on pieces of Leeds pottery with the date, 1811. Incidentally, it may be mentioned that it became the fashion to show engines on pieces of pottery. *See* page 64.

Some of the rails that Blenkinsop laid down were 6 feet in length with solid teeth, but most of them were of 3 feet long with hollow teeth. Each tooth measured 3in. by 2in. by 2½in. long, and the distance from centre to centre of teeth was 6in. The original 6 feet rail in the Leeds Museum was presented by Miss Maude, of Middleton, and there is a replica in the Science Museum, South Kensington, given by S. Denison & Sons Ltd., of Hunslet Foundry.

The pounding weight of the locomotive, combined with its tractive effort on the teeth, etc., caused breakages, and it was thus convenient to have a Foundry by the side of the line, from which new rails could be obtained. The Foundry was then owned by John Goddard and Daniel, the son of Titus Salt.

The teeth of the rails were of course made to suit the cog-wheel on the locomotive which Murray designed, and it was 38¼in. diameter with 20 teeth. It was driven by connecting rods from pistons of two vertical cylinders, so arranged that when one piston was in midstroke and full steam pressure, the other was at end of its stroke.

The locomotives ran for many years and the coal was delivered at the coal staithe, Great Wilson Street, at 7/- per ton for best quality and 4/- for an inferior quality.

One boiler burst, and Geo. Stephenson, in giving evidence before a Committee of the House of Commons about it, said :

> " The driver had been in liquor, and had put a considerable load on the safety valve, so that upon going forward the engine blew up and the man was killed. But he added, that if proper precautions had been used with that boiler the accident could not have happened."

By 1835, horse feed was again plentiful and the last locomotive being worn out, it was taken off and for some years was exhibited in a shed at Belle Isle, Middleton.

In 1862, the Middleton estates of the Brandling family being then in Chancery, were offered for sale at the " Scarborough Hotel," Leeds, and purchased by the Mauds and the Nicholsons. About this time the cast-iron rails, rack and plain, were purchased by Richard Kilburn, who melted them down in the same " air furnace " at the Foundry where they had been made about half a century before.

The track was then changed to an ordinary railway with wrought iron rails, and one of the tank locomotives was called " The Blenkinsop." As it ran at considerable speed through areas where there were now many houses, people complained about the unfenced line.

Agitation came to a head when the Colliery Company began to build a branch line over another part of Hunslet Moor to cross the highway in another place and join with the Midland Railway. John de Morgan, son of a contractor who built Morley tunnel, organised a meeting on the Moor at which

30,000 people were present, and some railway rails were removed.

The authorities interfered, and eventually, when the agitation quietened down, Morgan went to America and became Editor of a paper. In 1877 William Emsley, a solicitor of Leeds, caused copies of the original Acts of Parliament to be printed in pamphlet form.

Eventually the line was straightened and securely fenced, and being still in regular use, it is the oldest railway in the world. For the conveyance of coal it has a continuous history of 170 years. *See* also pages 108 to 110.

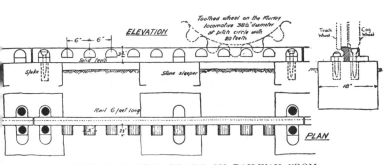

RACK RAIL USED IN 1812 ON RAILWAY FROM
MIDDLETON COLLIERY TO HUNSLET, LEEDS.
Rails designed by John Blenkinsop—Locomotives designed by Matthew Murray.

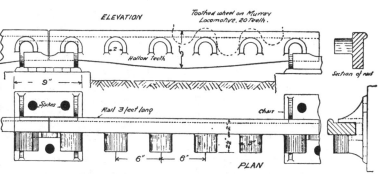

RACK RAIL USED IN 1813 ON LINE CONNECTING
KENTON & COXLODGE COLLIERIES TO WALKER-ON-TYNE.

BLENKINSOP'S PATENT FOR RACK RAILS.

The following are extracts from Patent No. 3431 of 1811, taken out by John Blenkinsop, aided by John Straker :—

Firstly, I do construct, place and fix in and upon the ground or road over which such conveyance is to be made a toothed rack or longitudinal piece of cast iron or other fit material, having the teeth or protuberances or other parts of the nature of teeth standing either upwards or downwards or sideways in any required position, and I do continue or prolong the said toothed rack or longitudinal piece by the addition of duly placing and fixing of other like pieces, all along or as far as may be required upon the said ground or road.

After stating that the carriage to bear or convey goods along the said ground or road must have a wheel having teeth or protuberances, he goes on to say :—

Thirdly. I do cause the said wheel to revolve and drive the carriage along by the application of any such well-known power or first-mover as can be placed upon and carried along with the said carriage, and I do declare that a steam engine is greatly to be preferred to any other first mover, and further that I do connect the said wheel with the first mover by crank assisted by a fly, or any other of the several methods of connection which are well known and easy to be applied and used by any competent mechanic.

It is obvious from his mention of " a fly," by which he means a flywheel, that Blenkinsop's idea of the engine was that it would have a single cylinder like those made by Trevithick. It could not have been a commercial success as was the locomotive built by Murray.

It is annoying that the mistake of giving credit to Blenkinsop for the locomotive designed and built by Murray should still persist. In earlier times the facts were known, for example, William Whitehead, of Wolverhampton, who saw the locomotive in 1824, wrote to the paper :—

" With a few more youngsters I took a stroll from York Road, Leeds, to Hunslet Pottery. Whilst there, I was greatly puzzled to see loaded coal waggons pass by with-out horses. I afterwards learned that the locomotive I then saw was built by a Mr. Murray, on an order by Mr. Blenkinsop."

Only last year the writer had to send a correction to a Leeds newspaper because the locomotive had been again called a Blenkinsop.

POPULAR FALLACIES.

For many years the public has nursed two fallacies regarding steam engineering, one being that the utilisation of steam in an engine was due to James Watt, and that he got the idea when as a boy he watched steam lift the lid of a kettle.

As a matter of fact his introduction to steam power came when, as an instrument maker in Glasgow, he was asked to repair a model of a Newcomen steam engine.

He saw a way of improving it by having a separate condenser, and took out a patent and joined Matthew Boulton, of Birmingham, to make steam engines. The firm supplied an engine to Whitbread & Co., in 1785, and fixed on it a plate with the following inscription, which is from a copy owned by the late Prof. C. Unwin, F.R.S. :—

> " The steam-engine was first invented by the most noble Edward Somerset, Marquis and Earl of Worcester, 1655. Improved by Captain Thomas Savery, 1699. Further improved by Thomas Newcomen and John Cawley, ingenious mechanicians, of Dartmouth, in the County of Devon, 1712.
>
> " The expense in fuel reduced one-half by Mr. John Smeaton, F.R.S., of Austhorpe, in the County of York, civil engineer, 1768.
>
> " Greatly enlarged in its powers and uses, and brought to its present state of perfection by Mr. James Watt, F.R.S., of the City of Glasgow, who received His Majesty's patent, 1769, and in conjunction with Mr. Matthew Bolton, of Birmingham, F.R.S., obtained an Act for the exclusive privilege for twenty-five years, commencing 1775.
>
> " They erected this engine, which performs the work of thirty-five horses, 1785. But in the course of ten years' use, more power being found necessary, this engine was altered to make it equal to the work of seventy horses, 1795."
>
> —See " Popular Fallacies Explained and Corrected," 3rd ed., by A. S. E. Ackerman, B.Sc. (Eng.), Lond., A.M.Inst.C.E.

The inscription mentions a fact that deserves to be better known, namely, that John Smeaton, of Leeds, improved the efficiency of steam engines by fifty per cent. before James Watt began to make engines.

The other fallacy is that George Stephenson's " Rocket," built in 1829, was the first successful locomotive, and that the Stockton and Darlington line opened in 1825 was the first successful railway.

To quote again from the above-mentioned book :—

> Smiles states that the first locomotive steam engine was built at Paris by the French engineer Cugnot, the model of it having been made in 1763. In 1772 an American, Oliver Evans, invented a steam carriage. In 1784, William Symington was at work in Scotland on the same subject ; and in the same year William Murdock, friend and assistant of Watt, constructed his model of a locomotive. Then came Richard Trevithick, Murdock's pupil. Stephenson carefully studied everything that was being done in the matter of steam locomotion in his time and before, and then set to and completed his first locomotive in 1814.

The writer drew Mr. Ackerman's attention to the following : *First,* that in 1811 Matthew Murray built the first locomotive, with two cylinders and having cranks set at right angles so that it could start the load from any position; *Second,* it began to run commercially on the Leeds to Middleton Railway line on the 12th of August, 1812; *Third,* that four such locomotives ran regularly for many years, the last being taken off in 1835.

Mr. Ackerman has been good enough to say :—

> Your claims for Matthew Murray in connection with the locomotive are entirely new to me, and therefore I cannot either confirm or correct what you say. But I will certainly look into the matter, and knowing the care you take in these matters I expect to find that what you say is correct. If I find that it is so, then undoubtedly I will add a paragraph to my article on the invention of the locomotive in the next edition of " Popular Fallacies."

It is necessary to keep reiterating facts about Murray and his work, because when a mistake once gets into print it is difficult to stop it spreading.

––––––––––

We devote vast departments of government, great agencies of commerce and industry, science and invention, to decreasing the hours of work ; but we devote comparatively little to improving the hours of recreation.—Herbert C. Hoover.

Scientists, engineers, chemists, and especially electrical engineers, have completely changed the art of living, even as compared with the last century, and therefore the men who are required for executive control should have the scientific, forward-looking minds which arrive at decisions by reasoning and research. For a long time the classically trained have been out of their depth, and as scientific and engineering progress speeds up, they get further behind every day.—E. Kilburn Scott.

LONGEVITY OF MURRAY'S ENGINES.

There have been many proofs of the excellence of design and long life of Murray's engines, for example, one supplied in 1813 to Water Hall Mills, Holbeck, ran continuously until 1885.

Another made about the same time for Carlton Mills, Leeds, ran until the early years of the war.

At North's Chemical Works in Saynor Road, in Hunslet, there was an old beam engine by Fenton, Murray & Wood which ran until the early nineties.

In the Locomotive Repair Works at King's Cross Station, London, an ancient beam engine has been working since 1848. It was bought second-hand from an old saw-mill in Agartown, near-by.

F. Haines, one of the workmen who was in the repair shop for over 65 years, remembers his father saying that he had seen the engine running at the saw-mills.

A comparison of the details of design with the illustration of an engine by Fenton, Murray & Wood, published in 1815, indicates that the engine was originally of Murray's design. The beam is supported by columns.

The engine still runs six days a week, sometimes Sundays —and all night in case of an urgent repair. It is probably the only engine in the world that is still running after being in use for over a century.

About 40 years ago a new cylinder with valve-gear was fitted, and this is 18-inch in diameter by 4ft. stroke. The engine now works with steam at a pressure of 110lbs. per sq. inch, and the speed is about 30 revolutions a minute.

Some day this repair shop will be driven by electric motors, and the engine will not be required. When that occurs it is to be hoped that the Railway Company will make arrangements to preserve the engine.

A beam engine of Murray's design, but built after his death, was running at the linen mills of Messrs. T. R. Leuty & Co., of Armley, for over eighty years.

STEAM POWER IN EVOLUTION.

The following article, from the pen of Mr. David Brownlie, appeared in the " Engineering and Boiler House Review " of September, 1927.

It gives the sequence of improvements very fairly, but unfortunately makes the old mistake about John Blenkinsop designing a locomotive. It is one of the objects of this book to correct that error.

Trevithick's high pressure " Cornish " boiler made the steam locomotive—quite impossible under wagon boiler conditions of 5-6 lbs. pressure—a practical proposition because of the reduction in boiler weight. Nothing will, of course, ever kill the idea that George Stephenson invented the steam locomotive, and that the Stockton and Darlington Railway opened in 1825, was the first in the world. The idea belongs to the same category as the popular impression that James Watt was the first man to use steam at all, because he watched a kettle boil when he was a boy, or that Arkwright invented cotton spinning.

The essential facts, on present knowledge, are that the first locomotive or vehicle driven by steam was that of Cugnot, in France in 1769, which ran on the roads but could only keep in motion for a short time on a level road, probably much less than ten minutes, because the boiler was not powerful enough. The subject subsequently attracted the attention of many inventors. For example, William Murdoch, in 1786, when resident in Cornwall, made some excellent working models of steam locomotives, but was compelled to abandon the work by his employers, Boulton and Watt.

Trevithick also constructed working models in 1797, after which, because of the perfection of the high pressure " Cornish " boiler, he concentrated specially on the subject, and in 1801 constructed a very large locomotive at Camborne, in Cornwall, first driving it on the roads on Christmas Eve, 24th December, 1801. Then, in 1802, Trevithick constructed a still larger locomotive, or steam carriage, and actually drove this about the streets of London. In 1804 he built the first railway in the world, at Merthyr Tydfil, running between Penydaran and Quaker's Yard, a distance of ten miles, hauling 25 tons of coal or other material, along with stray passengers, at a speed of four miles per hour.

In the same year he sent one of his locomotives to the Wylam Colliery in Northumberland, and it was from this that the whole of the work of the north-east coast originated, with which especially are associated William Chapman, John Blenkinsop, Timothy Hackwork, William Hedley, who constructed the well-known " Puffing Billy," and Jonathan Foster.

There were many locomotives in operation on the north-east coast before 1825, and all that George Stephenson did—with the very material assistance of Ralph Dodds and William Losh—was to improve existing designs, his " Locomotion No. 1," which ran on the Stockton and Darlington Railway, being essentially a mixture of Blenkinsop's and Hedley's locomotives.

It may be stated also that in 1808 Trevithick constructed a very remarkable locomotive known as the " Catch-me-who-can," which weighed 10 tons and ran at 12 miles per hour. He drove this in London, on a circular track in a field, charging 1s. per head. Hundreds of people rode upon it. Curiously enough, this track is to-day covered by Euston Station.

It may be added also that Richard Trevithick is not only the chief inventor of the steam locomotive, but he also originated many of its accessories, including the exhaust blast at the base of the funnel, the hand bellows to start up the fire before steam was raised, and the modern rail and flanged wheels. All these, including Trevithick's " Cornish " boilers, were used by George Stephenson in his locomotives.

James Watt, however, steadfastly regarded 6 lb. pressure as the proper limit for steam practice, ridiculing Trevithick and the steam locomotive ; while the firm of Boulton and Watt did their utmost, and very nearly succeeded, to get a Bill passed through the House of Commons to make the use of steam over 6 lb. pressure illegal, because, of course, the Watt engine and boiler combination was not suitable for much higher figures. Eventually, about 1820, Boulton and Watt were compelled to go to 10 lb. pressure. They also added an internal flue to the " wagon " boiler, but it was too late, and the main centre for steam engine and boiler construction on Trevithick's high pressure principles was transferred to Lancashire, while the firm of Boulton and Watt, after the death of the two founders, went out of existence.

The " Cornish " boiler continued to grow in size and pressure as the demands for steam and power increased. Eventually it was not possible to obtain the desired rate of evaporation with one furnace tube, so that two were added, side by side, forming the "Lancashire" boiler, so called because this development commenced in Lancashire, and was almost entirely due to the demands of the textile industries. In general it may be said that a typical steam boiler and power plant in a very up-to-date Lancashire cotton mill, about 1827, included a beam engine of 29-50 h.p. and a steam generation plant with " Cornish " or plain externally-fierd cylindrical boilers operating at 20-30 lb. pressure. The boiler was almost certainly hand fired, with the very hot waste combustion gases passing direct into the chimney.

" Matthew Murray, who, with exceptional skill and high reputation, founded the locomotive industry in Leeds, lies in a neglected and half-forgotten grave in Holbeck Churchyard, while the memory of the man who became so jealous of him that his firm bought land next to Murray's works with the object of preventing extensions, is commemorated in City Square."—W. J. Barker.

MILL DRIVING BY STEAM POWER.

When John Marshall removed his mill from Adel to Holbeck and Fenton, Murray & Wood moved their works from Mill Green to Water Lane, they both did so because of circumstances brought about by the coming of the steam engine.

Marshall's first mill at Adel used natural water power, and when, owing to Murray's improvements to his flax machinery, he found he wanted more power than Adel Beck could give, he moved to Holbeck in order to take advantage of steam power raised with coal from neighbouring pits at Beeston.

The engine was of the very early Savery type and as it could not drive the line shafting direct, it was used to pump water on to the buckets of an overshot water wheel. It is of interest to note that this method of obtaining *artificial water* power was sometimes used where there was also *natural* water power, the water being pumped from the tail-race to the head-race.

When Murray started his first works at Mill Green he was doubtless attracted there by natural water power that had been developed by Holbeck Beck. Directly he had made a steam engine to drive his own works the firm moved to a much more convenient site in Water Lane. It was thus near the Aire and Calder Canal and Leeds and Liverpool Canal.

In a Presidential address before the Newcomen Society, Mr. Rhys Jenkins traced the developments of mill driving from water power to steam power in the following interesting way :—

> The introduction of the water wheel was not due to any desire for economising human labour (that was cheap enough) ; it had for its object the performance of operations that could not adequately or conveniently be done by men ; for instance, the working of the great bellows and hammers in ironworks, and stamping mills for fulling, papermaking, etc.
>
> An increased number of wage-earners and the consumption of larger amounts of raw material and fuel, would alone tend to drive out the small man and to place an industry, previously in his hands, in the hands of a capitalist employer. Still more perhaps would the cost of the machinery, the water wheel with its pond and leat and the water rights, tend to do this.
>
> In this country the power available at any given spot on a stream is quite limited, so that although larger establishments become pos-

sible, the extent of the increase was limited. From one point of view, this was a good thing—it was against the growth of large industrial areas, and works were scattered along the banks of the streams all over the country ; but to the manufacturer it was a great drawback. If his product was such as to demand more than one operation, he might have to cart his half-manufactured goods from one mill to another, some miles away. If his business was expanding, his only course was to set up another mill with its own overhead charges and other sources of loss. Then the seasons imposed another drawback ; in winter the streams were locked by frost, in summer the water supply failed—in either case the mill was rendered inoperative.

All this was changed by the introduction of the steam engine. It became possible to concentrate all the processes of manufacture in one establishment and to carry on the work throughout the year. The manufacturer, it is true, had to pay for these advantages, the coal bill became a permanent charge on his undertaking. The change was at first a gradual one ; it originated with the desire of the owners of water mills to obviate the stoppages of work due to insufficiency of water in the summer months.

The Newcomen or atmospheric pumping engine was applied to return the water from the tail-race to the head of the water wheel. The plan proved advantageous, and as time went on more and more reliance was placed on the engine, until it became the essential element of the combination, and the water wheel merely a means for producing rotary motion. Then the question arose whether a mechanism could not be contrived to convert the to-and-fro motion of the engine into one of rotation, and thus to dispense with the water wheel. The problem was solved and the rotative steam engine was produced, and at a critical moment. It came into use with amazing rapidity.

Several people appear to have proposed changing the oscillating motion of the beam to the rotatory motion of a fly wheel shaft by means of a crank and connecting rod. The idea of a crank was old, but James Pickard, a button maker, obtained in 1780 a patent for applying it to steam engines.

This caused Watt to use the " sun and planet " motion, and Murray to use his cycloidal straight line motion until Pickard's crank patent expired. Watt introduced the parallel motion in 1784. His principal patent expired in 1800.

Jonathan Hornblower introduced compounding in 1781. In 1802 Richard Trevithick made a semi-portable engine, using steam at higher pressures. In 1804 Arthur Woolf demonstrated the superior economy of an engine in which steam was expanded over six times.

" FATHER OF ENGINEERING."

In the Parliamentary Poll list of 1807, Matthew Murray was the only voter in Leeds who was given the title of " engineer," the rest being called millwrights, ironfounders, and blacksmiths. It is quite proper to call him " Father of Leeds Engineering," because he was the first in Leeds to build flax spinning machiney; carding engines; pumping engines; stationary engines for mills; air pumps; locomotives; planing machines; cylinder boring mills; hydraulic testing machines; pressure gauges; and gas making plant.

A better title, however, would be that of " A Pioneer of Mechanical Engineering." His influence went much further afield than Leeds, and textile machines, stationary engines, machine tools, etc., were sent to all parts of the country and many places abroad. He also possessed in a high degree, that attribute of real genius, namely, a truly liberal mind.

Men he trained went about as erectors of machinery and engines, carrying the traditional skill in craftsmanship and design they had directly and indirectly imbibed by association with the great engineer. Some of his apprentices started businesses of their own, as for example, Charles Gascoigne Maclea, and Joseph Ogdin March, who married his daughters.

The Butlers, who built Kirkstall Forge, were with Fenton, Murray & Wood, and Benjamin Hick (the name Hick is on one of Murray's patent specifications), went to Bolton in Lancashire. He was with Rothwell, Hick & Rothwell, and later he started Benjamin Hick & Sons, now Hick, Hargreaves & Co., Ltd. At one time their engines drove the bulk of the spindles in Lancashire cotton mills.

Of those who went abroad, the most distinguished was Samuel Owen, who represented Fenton, Murray & Wood in Sweden from 1804 for several years, and then started a works in Stockholm to build engines, steamships, etc. In 1816, he fitted an engine and boiler to a vessel with paddle wheels. This was the first passenger steam boat built on the Continent and the forerunner of many that he built. There is a monument to his honour in the Public Square of Stockholm, and a Samuel Owen Scholarship for Engineering at the University.

PLATE 6.

MODEL OF LOCOMOTIVE AS MADE BY MURRAY, 1812. *In the Science Museum Collection, South Kensington.*

PLATE 7.

Picture of Murray's Locomotive Pushing and Pulling Waggons of Coal.
This view is taken near the present coal staithe in Great Wilson Street.
Christ Church is in the background.

[*Block lent by Meccano Ltd.*

THE MURRAY TRADITION.

After Murray's death, the firm was carried on as Fenton, Murray & Jackson, and with the creative impulse he had given, acting like a flywheel, the new firm carried out much important work. They specialised on locomotives, and many young engineers who passed through the works became eminent as locomotive superintendents.

One was Richard Peacock who, when only 21 years of age, was locomotive superintendent of the Leeds and Selby Railway. In 1854 he joined Charles Beyer in starting the famous locomotive firm of Beyer, Peacock & Co., of Gorton Foundry, Manchester, and he thus helped to make engineering history in Lancashire. Incidentally they introduced artistic lines into locomotive design.

Murray Jackson, who was a son of the partner, took the Murray tradition to Switzerland, where he was manager of the celebrated engineering works of Eyscher, Wyss & Co., of Zurich. Later he became chief engineer of the Royal Danube Navigation, and received honours from the Emperor.

John Chester Craven left the Round Foundry in 1873, and after being with the firm of Todd, Kitson & Laird and at the Railway Foundry in Hunslet, he placed his experience at the disposal of the famous London firm of Maudslays. In 1874, he became locomotive superintendent of the London, Brighton & South Coast Railway, and after holding it for 22 years, started the Victoria Docks Engine Works.

Luke Longbottom, son of the engineer who erected the pumping engine at Charlesworth's Colliery, Rothwell, in 1776, went to Kitson, Thompson & Hewitson in 1843, when he left the Round Foundry. He was later locomotive superintendent of the North Staffordshire Railway.

Another famous engineer of that time, also trained at the Railway Foundry, was David Joy, inventor of the Joy radial valve gear, which became practically universal on all large locomotives.

The name Kitson occurs several times in the above, and it will thus be seen how the Murray tradition became merged in the locomotive building firm started in 1837 by Todd, an engineer, and James Kitson and Laird, business men, to build locomotives, thus carrying on a branch of engineering initiated by Murray.

Todd, Kitson & Laird built several locomotives for the Liverpool and Manchester Railway, at the Railway Foundry, which was afterwards owned by E. B. Wilson & Co., connections of the well known Hull firm of ship owners.

In 1839 Todd joined Shepherd, whilst Kitson and Laird started another works which became the present Airedale Foundry, and the firm was Kitson, Thompson & Hewitson for about 20 years. The Kitson family became owners in 1865. (See *Engineer,* Nov. 23, 1923).

In 1862, Smith, Beacock & Tannett gave the name "Victoria Foundry" to Murray's old works, and until 1894 were occupied in making machine tools, a line of activity that Murray had initiated at the beginning of the century.

There is not space to give the names of scores of engineers who received their training at Smith, Beacock & Tannett, but the Brothers Krupp should be mentioned, because of the influence these men had on the history of Europe. Alfred Krupp was one of many continental buyers of machine tools from Smith, Beacock & Tannett, and was so impressed with the work done by the Holbeck firm that his sons came there for practical training.

One who was with the firm, called Pflaum, settled in Leeds and some years ago his descendant negotiated the sale of the Round Foundry site to present owners.

INSTRUCTORS OF ENGINEERING.

In the seventies, Smith, Beacock & Tannett had a foreman in the Smith's shop called Cryer, who was afterwards works manager. His son, Thomas, a draughtsman with the firm, started teaching mechanical drawing and applied mechanics in Holbeck Mechanics' Institute.

In 1884 he went to the Manchester Technical School to take charge of the classes in Mechanical Drawing and Applied Mechanics. Along with H. G. Jordan he published books on "Mechanical Drawing" and "Applied Mechanics," which contained examples of engineering practice as developed in Leeds. They were the pioneer books of the period.

After teaching at the Leeds Technical School and other places, Arthur Cryer became County Council lecturer in engineering in Glamorganshire and brought out books on " Engineering Drawing " and " Mining Drawing."

Other members of the drawing office staff of Smith, Beacock & Tannett who became teachers of mechanical drawing and engineering were George Oldfield; Thomas Tannett Heaton; Herbert Tannett Heaton; W. J. Robinson; H. Sutcliffe Myers; Thomas Jackson; Percy Bentley and Ernest Newell.

Thomas Jackson, M.I.Mech.E., succeeded Thomas Cryer at various Institutes and became Lecturer in Engineering under Prof. Archibald Barr at the Yorkshire College. Amongst those who studied under him and thus imbibed something of the Murray tradition were Prof. Ernest Wilson, M.Inst.E.E., of King's College, London; Samuel Rhodes, M.I.Mech.E., first principal of engineering at the Gordon College in the Soudan, and the writer who, when in charge of electrical engineering at the University, Sydney, passed on ideas in design, first learnt in Leeds.

In the eighties whilst Ernest Newell, M.I.Mech.E., was in the drawing office of Smith, Beacock & Tannett, and he also taught in the evenings. He is now the principal maker of cement machinery in the country and has a large works at Misterton.

Another very successful instructor who received his training in Leeds was Wilfred Lineham, who was in the drawing office of John Fowler & Co. Ltd.

He went to the Rutherford Technical College, Newcastle-on-Tyne, where for eight years he was lecturer on engineering subjects, and then to the Goldsmiths Technical Institute, London, where he brought out a text book on Mechanical Engineering known as "The Lineham," that has run through nearly forty editions. He did very useful work during the war in the production of extensively accurate gauges. He was an artist and had pictures hung in the Royal Academy.

In their teaching and books these Leeds engineers naturally used examples and details of design with which they were most familiar, and thus the tradition of mechanical design started by Murray has been spread abroad.

ELECTION POLL LIST OF 1807.

The candidates were Mr. Wilberforce (Independent), Lord Lascelles (Tory), and Lord Milton (Whig). The Poll was taken at York, and Lord Milton was at the head. The following is abstracted from the list :—

	Indt.	Tory.	Whig.
LEEDS.—Joseph Cawood, Ironfounder	1	1	
James Fenton, Ironfounder			1
John Houseman, Millwright	1	1	
Wm. Lister, Ironfounder (Bramley)	1	1	
John Nichols, Millwright			1
Wm. Labron, Ironmonger		1	
John Reynolds, Ironmonger		1	1
Joseph Shaw, Ironmonger		1	1
BRAMLEY.—Charles Lord, Millwright			1
CHURWELL.—John Bramah, Millwright			1
HOLBECK.—Matthew Mirror Engineer			1
John Sturges, Ironmonger	1	1	
David Wood, Ironfounder			1
HUNSLET.—Thomas Craven, Blacksmith		1	
John Gothard, Ironfounder	1		1
Wm. Gothard, Ironfounder			1
John Jubb, Millwright			1
Topham Roberts, Blacksmith			1
—— Salt, Ironfounder	1		
Robert Tidswell, Ironmonger		1	
David Trickett, Millwright	1		1
KIRKSTALL.—George Beecroft, Ironmaster	1	1	
Thomas Butler, Ironmonger	1	1	
John Butler, Ironmonger	1	1	
MORLEY.—John Hartley, Millwright			1
WORTLEY.—Thos. Bedford, Machine Maker	1	1	
Henry Firth, Blacksmith	1		
Thos. Knapton, Blacksmith	1		1
BRIGHOUSE.—George Hoyle, Millwright			1
Beny Pinder, Blacksmith			1
James Pinder, Blacksmith	1		1
DEWSBURY.—John Brook, Blacksmith			1
John Wormald, Blacksmith			1
ELLAND.—Thos. Mann, Machine Maker			1
HALIFAX.—Thos. Bradley, engineer		1	
John Emmett, Ironfounder	1	1	

Matthew Murray's name is mis-spelt " Mirror," that being the way the name was pronounced in Holbeck and is still pronounced by old residents.

He is the only voter of Leeds entitled "Engineer," his three partners, James Fenton, Wm. Lister, of Bramley, and David Wood, of Holbeck, being called " Ironfounders."

John Bramah, the millwright, of Churwell, was probably a relative of the inventor of the hydraulic press. John and William Gothard were at Hunslet Foundry, and Salt, the ironfounder, was a partner and father of Sir Titus Salt.

SIMON GOODRICH (b. 1773, d. 1847).

He started as draughtsman in the office of Sir Samuel Bentham, Inspector General of Naval Works, London, and rose to the position of Engineer and Mechanist to the Navy Board, his duties being concerned with the mechanical equipment of vessels and particularly with Government yards.

He gave Fenton, Murray & Wood very many orders for the dockyards of Chatham, Portsmouth and Plymouth, and was on such friendly terms with Matthew Murray, that he always stayed at Holbeck Lodge when visiting Leeds.

The Science Museum, South Kensington, has a set of his Journals from 1800 to 1831, and they are a very valuable record of engineering work at the beginning of the last century. There are many references to Murray's firm, including estimates for engines and hydraulic machinery, and in one case prices of apparatus for coal gas as given by Murray.

The collection includes a number of finely executed drawings supplied by Fenton, Murray & Wood, one having the initials B.H., which stands for Benjamin Hick, who founded Benjamin Hick & Sons, which later became Hick, Hargreaves & Co. Ltd., of Bolton.

There is a coloured drawing of the hydraulic press for which Murray obtained a patent in 1814, and it is of interest to note that it was Murray's original copy, and that he asked Goodrich to return it. Fortunately he did not do so, for otherwise it would have been destroyed with other drawings, etc., when the Round building was burnt in 1872.

Goodrich gives some particulars of Benyon's flax mill which was built about 1805 in Mill Street, Meadow Lane, and he says that it was on the same plan as Marshall's, they having dissolved partnership.

> The mill had arched floors built between cast iron girders which were supported on cast iron pillars placed about 10 feet apart. The mill was driven by a fifty horse power engine by Fenton & Co.

It is a mistake for those engaged in industries of the North to send their sons to Oxford or Cambridge because of the danger of them being educated " out of line." If they must go to a University let it be to one near at hand, where the training is at least in keeping with and in touch with industrial life and conditions.—E.K.S.

RICHARD TREVITHICK (B. 1771, D. 1833).

He was a great genius and saw the future possibilities of steam power more clearly than Watt or any others of the time, and especially the advantages of having high pressures. He introduced the Cornish boiler having a single flue, the fore-runner of Lancashire boilers with two flues, and these are still standard types.

Boulton & Watt opposed all Trevithick's attempts to introduce high pressure steam, and even went the length of trying to get a Bill through Parliament to prevent the use of steam pressures above 6 lbs. per sq. inch.

Trevithick was the first to make a locomotive to haul loads on a railway and this he did at Penydarran in 1804. On the 2nd of March in that year he wrote to Davies Gilbert :—

> " We have tried the carriage with twenty-five tons of iron, and found we were more than a match for that weight. The steam is delivered into the chimney above the damper . . . it makes the draught much stronger by going up the chimney."

" Engineering," of 27th March, 1868, said of Trevithick that he—

> " Was the first to prove the sufficiency of the adhesion of the wheels to the rails for all purposes of traction on lines of ordinary gradient ; the first to make the return flue boiler ; the first to use steam jet in the chimney ; and the first to couple all the wheels of the engine."

In 1814 he became absorbed in a scheme for pumping mines in Peru, and nine of his engines made by Walker & Rastrick, of Stourbridge, were sent out to that country. He went to set them to work and remained about ten years. He lost everything during a political revolution.

In 1826 he met Robert, the son of George Stephenson, in South America, and travelled with him to New York, imparting information that helped the Stephensons to build improved locomotives, especially the " Rocket " of 1829.

When Trevithick arrived in England he was penniless, and a petition was presented to the Government on his behalf in 1828, but was disregarded. After a visit to Holland, he settled down in Dartford, and was engaged in the ironworks of John Hall when he died.

Some of Hall's workmen paid for the funeral and for men to watch the grave, because it was the time of "body snatching."

JOHN SMEATON (B. 1724, D. 1792).

It is usual to associate the name of Smeaton with the construction of lighthouses and other *civil* engineering work, but his work in *mechanical* engineering was equally important. It was mostly done in Leeds.

He was the first to delve into the underlying principles of the steam engine, and his scientific knowledge enabled him in 1768 to cut down the cost of fuel in engines by one half.

James Watt called him " Father Smeaton "; " his example and precepts have made us all engineers," and Robert Stephenson pronounced Smeaton to be " the engineer of the highest intellectual eminence that had yet appeared."

JAMES WATT, Junior (B. 1769, D. 1848).

He went to Paris to study in 1789, and being involved in the revolutionary movement, was at first in favour with the leaders. Later he was denounced before the Jacobin Club by Robespierre but escaped to Italy, and stayed until 1794.

He then became a partner in the Soho firm of Boulton & Watt, being principally stationed in London. He was in Leeds for some months in 1802. He is said to have given some assistance to Robert Fulton, the American, who built the " Clermont," the first steam boat on the Hudson River.

In 1817 he fitted the " Caledonia," a boat of 102 tons, with new engines and went to Holland and up the Rhine to Coblenz.

He died unmarried, being the last of the family of James Watt.

WILLIAM MURDOCH (B. 1754, D. 1839).

He was devoted to the interests of his employers, Boulton & Watt, and thus his fame has been somewhat overshadowed by them. He was very like Murray in being a practical engineer and having vivid imagination. He developed gas lighting.

James Watt was very much against any experiments with steam locomotion, and said he wished :—

" Murdoch could be brought to do as we do, to mind the business in hand, and let such as Symington & Sadler throw away their time and money hunting shadows."

MATTHEW MURRAY, Junr. (b. 1793, d. 1835).

The son was born in Holbeck, served an apprenticeship with his father's firm and went to Russia about 1821. He had an engineering business there, was married, and died in Moscow in 1835.

The following are extracts from three letters he wrote to his sister Mary, the first being from Tamloff, dated August 6th, 1822 :—

" I sent a Note to Mr. McLea three Months ago, which I have no doubt he has receiv'd long before this, as for writing to my father I really have nothing interesting to inform him of, as there is no likelyhood of any trade in the engine way from England whilst the Xchange continues so low. I am happy to hear he is well & hope full of business, I assure you I have my hands pretty full at present, but as the Song says (the Money comes *slowly* in), however I cannot complain—

" I have sold one half of my Estate, for which I have receiv'd nearly half the Sum I gave for the whole, & I have still with Mr. Davies a large Place with a deal of Buildings and my fine house as usual which comes now to be a very cheap purchase, tho' I assure you I did not buy it with any of the good Intentions you mention, but as a Home to go too, as my business requires me to be continually backwards and forwards in Moscow.

" Thy poor Jack has got a little thinner with knocking about, but I live well & was never better in Health. I begin to have some likeing for myself, therefore promise to send you my portrait in Miniature by the first opportunity, as it may be a long while before you see the Original.

" I don't now think Mr. Davies will be in England this year, he is engag'd building a new theatre in Moscow, the whole of the Boxes, Galleries, etc., are supported by Iron Brackets, without any Columns whatever to intercept the view, which is a great Improvement in my opinion, as that is the principal inconvenience in Drury lane & Covent Garden & all your theatres.

" In your next I should like to know if you have bought Mr. Wood's house, & what part of the world young David is in."

Young Murray appears to have depended on his sister to send him special clothes, for he writes :—

" I am much oblig'd by your kind present of the Waistcoats, which I assure you *fit exactly,* tho' I have not yet receiv'd them, and have wrote to Petersburg about them, also the Flannels my Mother has the Goodness to send.

" There is little going on here but Balls, Plays, Masquerades, Assembly's, &c. Madam Catalini has been singing here & will be here again in Feby.

" Will thank you to put in Hands and finish for me with all expedition 12 fine Ruffled Shirts, to be sent out with the engine I have order'd, the Money my Father will pay you on my Acct."

In another letter he refers to his grandmother, who was then living with the Murrays in Holbeck, and there is also a reference to his wife :—

" You ask about my house, I cannot better satisfy you than referring you to a plan I intend sending my father before I leave Mosco which will be soon. The domestic Concerns of my family are carry'd on & conducted by three old Soldiers and two Soldiers' wives. My only furnish'd room serves for bed room, compting house, dining room, etc., etc. My principal drawing or dining room is occupy'd for the present with Model Makers, my music room, study and intended bed rooms are also employ'd, some as Magazines and others with workmen ; should this arrangement not exactly suit your taste, I cannot at present help it, but it is nevertheless true. I am knocking down an old House as big as the round building, and am going to build some new ones, but not like the heiress of the pump I have Tenants before the Houses is begun. I will write to you again soon, as I am going to Vladimir where I shall have more time."

The last letter from Moscow, dated Jan. 21st (Feb. 2nd), 1827, is to Joseph Ogdin March :—

" I am perfectly pleas'd with all your proceedings, as well with your Partnership with my sister Mary as with my brother MacLea, & sincerely hope you will prosper in both. Your foundry seems to me to be very nice and convenient, and only wants what I am sorry to see you complain of, plenty of Work. I never hear what they are doing at the old Shop.

" Engines I get made at other Works and bring them to Mosco to dispose of. I keep a few Men to fix them, and have a small Shop, where I make whatever Machine comes to hand, or anything new that I think of ; I have had a run of logwood Machines.

" I have made the Drawings and Estimate for Government, for supplying the town of Mosco with Water, and Xpect the Order very soon, there will be two 24 H. Engines in one House, pumping thro' Iron Pipes into an elevated Iron Cistern, at the Distance of 3 Miles from the Engines, from which Cistern it is conducted by Iron pipes thro' the different Streets of Mosco, this is an immense Job and will take me about 3 Years. I have had some Opposition from Baird, but I hear he has dropt it, meeting no Encouragement.

" The time is coming when I ought to receive my £1,000, which I assure you will greatly assist me in this Job. My best love to Mary and my little new Relation."

The new relation was Mary Murray March, eldest sister of the late Miss Rosa March, who lived at Birch Grove House, Leeds.

LATER HISTORY OF THE WORKS.

Matthew Murray had three daughters and a son. Margaret, the eldest, married Richard Jackson, who became a partner, the firm being styled Fenton, Murray & Jackson.

The second daughter, Ann, married Charles Gascoigne Maclea, and Mary married Joseph Ogden March, and these sons-in-law started the firm of Maclea & March.

Fenton, Murray & Jackson supplied locomotives for the Liverpool and Manchester Railway, the Leeds and Selby Railway, the Great Western Railway, and the line between Paris and Versailles. Amongst other work they also supplied side lever marine engines for steamships plying between France and Constantinople.

In 1837 there was a strike and some of the best of the engineers left, and the firm ceased operations in 1843.

A number of leading employees combined together to carry on the business, and this was the first example of an industrial co-partnership. They took over most of the plant.

They were nicknamed " Forty Thieves," after a pantomime then running at the Theatre Royal, Hunslet Lane. From time to time various partners, Leach, Midgley, Pollard, Pybus, Spence, Walker, Whitehead, etc., were bought out by the others.

Eventually the place was owned by Smith, Beacock & Tannett, and this firm carried on until 1894, Joseph Craven being the last survivor. His memory is perpetuated by the Craven Engineering Scholarship at Leeds University.

The original round building from which the premises received the name " Round Foundry," was partly used for storing the drawings, patterns, etc., of Murray.

Part of the old wall of the building can be seen in the Ginnell that joins David Street and Marshall Street.

In 1862, Smith, Beacock & Tannett called the works the " Victoria Foundry," and this name, with the date, is on the large cast-iron posts at the entrance to premises now occupied by Messrs. Leech & Sinkinson, Ltd.

For many years the late Joseph Henry occupied one of the original foundries, and to-day several engineering firms, including Messrs. R. W. Crabtree & Sons, Ltd., printing machine makers, occupy parts of the site.

When the Murray family lived in Holbeck, they regularly attended Holbeck Church, occupying one of the large old-fashioned square pews. The family vault, at the extreme North end of the Churchyard, has over it a large obelisk of painted cast-iron, having the following inscriptions :—

In a vault underneath are deposited the remains of MATTHEW MURRAY, civil engineer, of Holbeck, who died the xx of Feburary, MDCCCXXVI., age LX years. Also of MARY, his wife, who died the XVIII. of December, MDCCCXXXVI., aged LXXII years.

Beneath this monument rest the remains of ANNE, the beloved wife of WILLIAM MURRAY JACKSON, engineer, of Holbeck. Born Octr. 6th, 1817. Died Septr. 4th, 1880. Also of the above-named WILLIAM MURRAY JACKSON, of Holbeck, grandson of the before-named MATTHEW MURRAY. Born Octr. 19th, 1816. Died April 6th, 1882.

Underneath this Monument are interred the remains of MARGARET, wife of RICHARD JACKSON, of Leeds, and daughter of MATTHEW and MARY MURRAY, of Holbeck. She died at Southampton the xxiii. of October, MDCCCXL., aged LIV. years. Also MATTHEW MURRAY JACKSON, mining engineer, youngest son of the said WILLIAM MURRAY and ANNE JACKSON, who died December 26th, 1901, in his 40th year.

Inside the Church there is a Tablet with the following inscription :—

Sacred to the memory of MATTHEW MURRAY, who died 20th of February, 1826, aged 60 years. Also of MARY, wife of the above, who died December 18th, 1836, aged 72. Also of MATTHEW, their son, engineer, who died at Moscow, July 22nd, 1835, aged 42 years. This Tablet is erected by their affectionate daughter, ANN MACLEA, of Leeds, 1837.

Another Tablet is to members of the March family :—

Sacred to the memory of JOSEPH OGDIN MARCH, who died 13th December, 1830, aged 11 months and 20 days. Also of ANN MACLEA, born February 24th, 1839, died February 21st, 1847. Also of MARY, mother of the above children, of Beech Grove House, Leeds, who died January 18th, 1864, aged 66 years. Also of JOSEPH OGDIN MARCH, of Beech Grove House, Leeds, born 17th February, 1799, and entered into rest 27th February, 1888.

MURRAY MEMORIAL TABLETS.

It is proposed to erect Bronze Memorial Tablets :—

(A) **On the site of the " Round Foundry," Holbeck.**
Messrs. R. W. Crabtree & Sons Ltd. agree to erect it.

(B) **In a central position in the City of Leeds.**
As prominently as the statue of James Watt.

(C) **At the Institution of Mechanical Engineers, London.**
Near to the Tablet to George Stephenson.

The first Tablets will have enamel lettering easily read from a distance, giving brief references to some of the following :—

In **1788** Matthew Murray started with John Marshall, flax spinner, of Leeds, and introduced into his mill many improvements, including " Sponge weights " and machinery for " Wet spinning." These revolutionised the flax trade, and Leeds became the leading centre for making flax machinery.

He greatly improved and standardised mill-gearing, textile machines and machine tools, and by personal example and training of others established " traditions " for well-proportioned designs and good craftsmanship.

In 1795 he commenced business with David Wood, and later James Fenton and William Lister were partners in the firm. The works of Fenton, Murray & Wood, in Holbeck, became widely known as " The Round Foundry," and were famous for making efficient mill engines. The premises were afterwards occupied by Smith, Beacock & Tannett, machine tool makers, and they handed on the Murray tradition.

He was responsible for many improvements in early steam engines, including the box or three-ported slide valve and a cycloidal straight line motion which was later employed for driving printing machines.

In 1811 he designed and made steam locomotives for the rack railway line, laid down by John Blenkinsop, between Leeds and Middleton. The four locomotives on this line were the first to be commercially successful on any railway. They were the first to have double cylinders for driving cranks set at right angles, and thus were able to start with full load. He was the first to use a spring safety valve and feed pump on a locomotive.

In 1816 he used the double cylinder arrangement for the first engine to be fitted into a Mississippi steam boat. He introduced a side lever type of engine, afterwards made by Fenton, Murray & Jackson and others, and used on steamships.

He made the first Cylinder boring mills with screw feed and early metal planing machines. He invented a double-acting Hydraulic press and the first indicator for measuring high hydraulic pressures. He made researches on strength of materials with a 1,000 tons chain cable testing machine.

His house in Holbeck, known locally as " Steam Hall," was the first to be centrally heated by steam. He pioneered in smoke consuming devices and made the first coal gas lighting plant for Leeds.

MEMORIALS TO CONTEMPORARY ENGINEERS.

JAMES WATT. Born at Greenock 1736, died at Handsworth, Birmingham, 1819.

Monument by Chantry in Westminster Abbey, with epitaph by Brougham.
Bust by Chantry in National Portrait Gallery, Edinburgh.
Portraits by Henning & Dawe in National Portrait Gallery, Edinburgh.
Portraits by C. E. de Breda and H. Howard in National Portrait Gallery.
Statues in the City Squares of Birmingham and of Leeds.
Statues in the Public Library, Greenock, and Handsworth Church.

WILLIAM MURDOCK. Born at Bellow Mill, Ayrshire, 1754; died in Birmingham, 1839.

Bust by D. W. Stevenson on the Wallace Monument at Stirling.
Murdock medal, awarded annually by the National Gas Institute.
Portrait in oil, by John Graham Gilbert, in Royal Society of Edinburgh.
Bust, by Chantry, in Handsworth Church, Birmingham.
Bust of Papworth, in City Art Gallery, Birmingham.

RICHARD TREVITHICK. Born at Illogan, Cornwall, 1771; died at Dartford, Kent, 1833.

Memorial window in the North Aisle of Westminster Abbey.
Trevithick Engineering Scholarship at Owen's College, Manchester.
Trevithick Medal awarded by the Institution of Civil Engineers.
Portrait by Linnell in the South Kensington Museum.
Bust in the Royal Institute of Cornwall.

GEORGE STEPHENSON. Born at Wylam, 1781; died at Chesterfield, 1848.

Memorial Hall at Chesterfield opened in 1879.
Medal at centenary of birth, held at Newcastle-on-Tyne.
Statue by Bailey outside Euston Station, London.
Statue by Gibson in St. George's Hall, Liverpool.
Statue by Lough, near High Level Bridge, Newcastle.
Two oil paintings by John Lucas in the Institution of Civil Engineers.
Painting by Pickersgill in National Portrait Gallery, London.

ROBERT STEPHENSON. Born at Willington Quay, 1803; died in London, 1859.

Buried in Westminster Abbey; bronze statue and brass, by Marochitti.
Portraits by Phillip and by Lucas in the Institution of Civil Engineers.
Portrait by G. Richmond engraved for Jeafferson's " Life."

All the above have biographical notices in the Dictionary of National Biography, and there is also a column to William Hedley, born 1779, died 1850, who, at Wylam in 1813, made the locomotive called " Puffing Billy." Timothy Hackworth, born 1786, died 1850, the blacksmith who helped Hedley, has no notice, but a statue has recently been erected at Shildon.

Monuments to Henry Maudsley in Woolwich Churchyard and to Matthew Murray in Holbeck Churchyard are of cast iron. The biographical notice to Matthew Murray in the Dictionary occupying two columns, is by R. B. Prosser.

Samuel Owen, the engineer trained by Murray, who went to Sweden in 1804 for the firm, has a public statue in Stockholm. There is an Owen Engineering Scholarship at the Swedish University.

CENTENARY SERVICE AND SERMON.

On Sunday morning, 21st February, 1926, a Centenary Service was held in Holbeck Church, attended by Councillor J. Arnott, the Lord Mayor, and members of the Institutions of the Civil, Mechanical, and Electrical Engineers; the Leeds branch of the Engineering and Allied Employers' Association; the Leeds Association of Engineers; the Amalgamated Engineers' Union; the Newcomen Society of London; also Messrs. Walter Fourness and W. J. Barker, who assisted the Incumbent in making the arrangements.

The Lessons were read by Lt.-Col. Kitson Clark and Councillor H. Briggs, and after the service there was a procession of clergy, choir and congregation and visitors to the Murray vault, where there was another short service.

The following is the sermon, which was preached by the Rev. R. J. Wood, Incumbent, from the text, " Seest thou a man diligent in his business? He shall stand before kings; he shall not stand before mean men."—Proverbs xxii., 29.

Of late years there has been an outcry (with which I find myself in complete sympathy) against a form of religion which would isolate God from the affairs of the work-a-day world. Men have been proclaiming that any form of Christianity which is true to the nature of God must concern itself with the industrial life, the artistic life, the political life, not only of our own Nation but of all Nations.

This modern reaction against the individualism of former days has its dangers, as for example the false conception that selfishness parading as patriotism, or selfishness parading as class consciousness, is anything other than selfishness. Selfishness, whether of an individual man or of a mass of men is equally contrary to the spirit of Christ. But at the bottom this modern tendency is a righteous protest against an attempt to exclude God from His lordship over every department of human life, and as such we may welcome it.

I have sometimes wondered whether in our insistence on God's interest in the social life of men, we are not now in danger of forgetting the equally vital fact that He is the Father of each one of us, and in particular that He is interested not only in our prayers and our morals, but in our home, our meals, our amusements, and (above all) in our work.

The most God-fearing among us are apt to exclude our daily work from the things with which we associate the thought of God. We recognise, of course, that God will expect us to be *honest* in our work ; but we do not often think of Him as expecting us to be *industrious*. Honesty in our work we conceive of as a moral quality ; diligence at our work we treat as a non-moral, or at any rate a non-religious thing.

94

It is all part of that dualism between the secular and the sacred which the Devil is ever forcing upon us. If he can rob God of six days of a man's week, he will gladly concede the seventh. But God is interested in the whole of life. It may be necessary for a preacher constantly to emphasise the duty of living in conscious fellowship with God, in prayer and worship, and it may be a convenience to him and to others if he speak of this as " developing the spiritual life " ; but at bottom it is an abuse of words.

Almost any human activity is a spiritual process. A man's whole personality is bound up in his bundle of life, and there is as much stupidity in the attempt to limit the sphere of God's interest to certain elements in man's nature, as in the parallel fallacy which recognises God's activity in the extraordinary, and is blind to His activity in the normal working of natural law.

Although, therefore, no service rendered to humanity can ever relieve a man of his duty of responding to the love of God by a life of conscious fellowship with Him in and through His Church, we must remember that it is God Himself Who fashions the mechanic as well as the mystic.

Marconi no less than Dante and St. Francis is the creature of the Holy Spirit. The life-work of Michael Faraday is no less prophetic than the life-work of John Wesley ; for each proclaims a thunderous " THUS SAITH THE LORD."

Here we touch one of the great points of contrast between the teaching of the Old Testament and the New. The Old regards the necessity of human toil as a proof of God's anger. It was the punishment inflicted by Him upon a disobedient race ;

> " Because thou hast hearkened unto the voice of thy wife, and hast eaten of the tree of which I commanded thee, saying, Thou shalt not eat of it, cursed is the ground for thy sake ; in toil shalt thou eat of it all the days of thy life. . . in the sweat of thy face thou shalt eat bread."

The pessimistic writer of the Book of Ecclesiastes finds human labour even more vain than human merriment ;

> " He that diggeth a pit shall fall into it; and whoso breaketh through a fence, a serpent shall bite him : whoso moveth stones shall be hurt therewith, and he that cleaveth wood is endangered thereby."

But a MS. discovered in Egypt not many years ago ascribes to our Lord a saying which, if indeed it be His, we must regard as the divine repudiation of so false a creed :

> " Lift the stone," He says, " and there shalt thou find ME ; cleave the wood, and I am there also "

The words are God's blessing upon all simple diligent human labour. I am not much concerned whether you accept them as genuine or not. The spirit of them breathes through the whole of the Gospel story. The Son of God Himself lived and worked as a carpenter in the country village, and when the day came for Him to gather around Him the little band of followers who were to form the nucleus from which should grow the Kingdom of Heaven, His choice fell upon men whose hands were hard and horny like His own.

" Is not this the carpenter?"—the question was asked in astonishment by His countrymen, as they listened to His words and saw the mighty works wrought by His hands; and we re-echo it to-day, but in a tone of triumph, for it reminds us that human labour has been consecrated for ever by the life of the Son of God.

The *purpose* which brings together a body of Christian men and women in the Lord's House on the Lord's Day must ever be the worship and praise of God, but the *occasion* which has brought this congregation together is the hundredth anniversary of the death of a great Holbeck man who, though his work may not yet have received the recognition which is its due, must rank amongst the great engineers of the world,— a man whose diligence in his business does indeed entitle him to stand before kings.

Matthew Murray was born in 1765. Leeds cannot claim the honour of being his birth-place, for the first years of his life were spent at Newcastle-on-Tyne, where he was apprenticed in a mechanic's shop. When he became a journeyman he worked at Stockton-on-Tees, but three years later, trade became slack, and he was driven to tramp to Leeds to find fresh work.

About 1789 he found employment with John Marshall at Adel, and in the following year began his career as an inventor in the fields of spinning-machinery and machine tools. His success brought him a gratuity of £20 and the post of foreman in the millwright's shop.

In 1795 he left Marshall's, to become a partner with James Fenton and David Wood, with their works in Mill Green, Holbeck. There Murray took charge of the engine-building. This is not the occasion to speak of the technical side of his work. As you heard just now, one who is eminently qualified to do so, will be lecturing to-morrow evening; but the veriest layman in matters engineering will recognise the debt of gratitude we owe to the man to whom is very largely due the actual *commercial* use of the locomotive steam engine.

There are few more interesting or romantic spots in Leeds to the student of the City's history than the old railway from Middleton to Great Wilson Street, built by John Blenkinsop for Matthew Murray's engines to drag their loads of coal over Hunslet Moor, almost into the centre of the town.

The three partners built a new works in Water Lane, and Murray built himself the house known as Holbeck Lodge, which still stands in the railway triangle. There he lived and worked until on 20th February, 1826, he died. He was buried just outside the walls of the old Church, which was taken down not very long after.

Twenty-five years later the municipal authorities brought about the closing of the Churchyard, unhappily without making any provision for

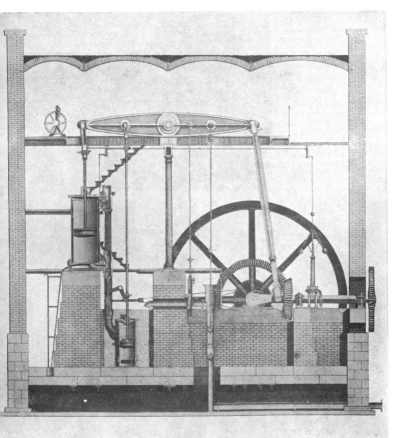

STEAM ENGINES
of improved Construction.
Designed and Executed by M. MURRAY Leeds.

PLATE 8.

MURRAY'S CENTRE COLUMN BEAM ENGINE.
(From a drawing in the possession of Mr. Ralph Murray Thompson.)

Built by
MATTHEW
MURRAY
about 1802
and heated
by steam.

HOLBECK
LODGE,

OR

" STEAM
HALL."

CAST IRON OBELISK IN HOLBECK CHURCHYARD.

PLATE 9.

the care and upkeep of it. Since that date most of the wealthier Church-folk have ceased to live in Holbeck, and it is quite impossible for the present generation to finance the upkeep of so large a piece of ground. But it is indeed a matter of regret to all of us that the grave of so famous a townsman, or indeed that the graves of the humblest of his contemporaries, should appear to be so neglected.

In view of the very prominent part played in the history of Holbeck Church by his relatives, one turns with interest to the records of the Church to discover what part, if any, Matthew Murray played in the Church life of his day. It was at first disappointing to find that no record appears of his taking any share in the official work of the Church. But a chance discovery of an old Poll-list of the Parliamentary Election of 1807 may afford an explanation of the absence of his name from the lists of Wardens and Vestrymen.

These were days when by one of the most tragic accidents in her history the Church of England had allowed herself to be identified with one political party, a tragic error, from the memory and the consequences of which the Church in Holbeck at any rate has never yet entirely recovered. In 1807 there were three candidates for the two seats in Leeds, Lord Milton, the Whig, Mr. Wilberforce, an Independent, and Lord Lascelles, the Tory.

Joshua Brooke, the Incumbent of Holbeck, plumped for Lascelles, and almost all the Holbeck people whose names appear in the Church lists voted Tory. " Matthew Mirror, engineer," voted Whig. Many turbulent years had yet to pass before men of all political beliefs could work together for the extension of God's Kingdom at the old Church of Holbeck.

But if all that I have been trying to say to-day be true, it would be a very partial and incomplete view of the life of God and of the life of man were we to think of the work of the priest or the Church-officer as being God's work, and that of the musician, the engineer or the brick-layer as being secular. " Every good gift and every perfect gift is from above, and cometh down from the Father."

The men who guide the destinies of our City; the architects who fashion its buildings; the engineers in charge of its lighting; the drivers of its trams; the sweepers of its streets are all the *employees* of God, Who is directly concerned with all the activities of His children.

God is the Father,—that is the great revelation concerning the nature of God made by Jesus Christ. Earlier writers had indeed used the name " Father " descriptively of God. But it is from our Lord pre-eminently that we learn that if our finite human minds are to get as near as may be to. the truth about the Nature and Being of God, we must think of Him in terms of perfect Fatherhood.

It is, of course, only an analogy, but it is an analogy which we may dare to press a very long way. God is the Father of Matthew Murray, as He is the Father of each one of us. And there are two joys which any father may know. One is the joy of a son's love; we may all of us render God that. The other is the joy of his son's ability, industry and integrity. " Seest thou a man diligent in his business? he shall stand before kings."

There are many who have so stood and have yet counted their greatest joy to have been the simple " Good lad !" of an earthly father's approval. The New Testament tells us that a like commendation (" Well done, good and faithful servant ") will be the greatest joy of those who in their days on earth have tried to use to the full the talents with which their Heavenly Father has endowed them.

The innermost secrets of the human heart God alone can read, and even if a hundred years had not passed we could but guess : But I dare to believe that God is proud of and pleased with His son, whose labours we commemorate to-day. And we Holbeck folk and you, his fellow-craftsmen, do well to thank God for Matthew Murray's labours and to share our Father's joy in his achievements.

Speaking at the graveside, Lt.-Col. Kitson Clark said he had been instructed by the Newcomen Society to deposit a wreath upon the monument of Matthew Murray, and in doing so he had to report that he was also directly representing the President of the Institution of Mechanical Engineers. There were also present representatives of the Institution of Electrical Engineers, and of the Institution of Civil Engineers, all brought together to bear testimony to the great reputation of the very great man whose grave they were facing.

Matthew Murray had many difficulties to overcome, and the world at large had so far failed fully to realise what it owed to his genius and his work. When the steam engine was first beginning to be made it was said that it could never be used for making flax ; but Murray had the courage of his opinions, and by his demonstration revolutionsed the industry.

Leeds had cause for pride in Murray's memory, and he hoped that the time would shortly come when others working in his steps would, by the elimination of smoke, make it possible for that sad, blackened spot to be one of sunlit beauty.

The Lord Mayor said it was a strange coincidence that, attending as he was in his official capacity, he was also a member of the Amalgamated Society of Engineers, and had served his time as an apprentice in the same town as Matthew Murray—Stockton-on-Tees. Engineers seemed to be present in force, for the Deputy-Lord Mayor was an official of the Locomotive Engineers' Union.

The work of Matthew Murray should be rightly remembered by Leeds with pride and satisfaction, but it should teach them also the lesson that hopes and aspirations were valueless without effort and courage to put them to the test. Like Col. Kitson Clark he, too, looked forward to the time when smoke would be eliminated and the city would be pure and fresh.

" One touch upon the lute of life well spent
Wakes the full orchestra omnipotent."

'WELL DONE, THOU GOOD AND FAITHFUL SERVANT."

By strenuous toil who may not, if he will,
   Unheeding every easy by-path lure,
   Some noble project of the mind mature,
Or win the guerdon of creative skill?
Such patient striving bold endeavour still
   Heartens to rich account, and shall ensure
   Content more precious, longer to endure
Than bed of luxury or pleasure's thrill.

What boots the transient boon of earthly gain?
   Through sacrifice and service, strain and stress,
   The steadfast soul attains true happiness,
With honour shining ever in its train,
And reads his name—a victor in the strife—
Writ large in the Remembrance Book of Life.

W. Herbert Scott.

Oh well to know what then we knew;
That firm resolve can win right through
The thorny maze, and reach the height
Of derring-do!

(From " The Lute of Life," published by T. Werner Laurie,
Ltd.).

CENTENARY LECTURE.

In connection with the Matthew Murray Centenary, a lecture was given in St. Matthew's Schoolroom, Holbeck, on 22nd February, 1926, by E. Kilburn Scott, A.M.Inst.C.E., of London, the chair being taken by Mr. C. Harding Churton, M.Inst.E.E., whose works occupy part of the old " Round Foundry." The lecture was illustrated by lantern slides; an original model of Murray's locomotive, and a length of rack rail three freet long, used by John Blenkinsop.

Technical details are given elsewhere in this book, and the following Notes are principally concerned with matters of local interest and of a more personal nature.

The lecture has been given at the University of Leeds and other places.

Some of those who wrote about Matthew Murray were not engineers, and they made a few mistakes which I think it well to try and rectify. They suggest, for example, that Murray lacked education, and by that doubtless meant some ability to read " a little Latin and less Greek " which passed for education in olden times.

As a matter of fact, Murray had just that particular knowledge which is considered the right kind of education to-day, and it was certainly useful, as it helped to make him one of the principal expert creative or constructive mechanical engineers of the " Industrial Age." His original letters show good penmanship and phrasing, and include apt quotations.

Various writers describe him as a whitesmith, a blacksmith, a machine-smith, and mechanic, none were quite right ; millwright would have been near the mark, because of his wide knowledge and adaptiveness. Engineer is the correct title, although some may say he only qualified for it when he began to make engines. When he died the words " Civil Engineer " were inscribed on his tomb, and it was quite justified, for he could have held his own with full members of the Institutions of Civil and Mechanical Engineers.

That he was accustomed to making engineering calculations is shown in a letter to Simon Goodrich, engineer and mechanist to the Navy Board, in which Murray gives the steps for calculating the horse power of an engine and works it out to three places of decimals. It was very characteristic of him to pass on such information to another engineer.

I doubt whether at the time half a dozen people in Leeds could have made such a calculation accurately, or even to-day that number of people could do it in, say, an average audience at the Town Hall. To suggest that Murray was not educated or " mathematical " is quite wrong.

Some books and articles have described his coming to Leeds when work was slack at Stockton-on-Tees, as if he made a " shot in the dark," but that again is wrong, because he had already had experience with flax machines, and he knew that it would be valuable to John Marshall, who was engaged in flax spinning at Scotland Mill, near the Seven Arches. The mill was driven by water power from Adel Beck.

Marshall was then paying a royalty to J. Kendrow, of Darlington, for certain flax machines, and it is significant that after he had adopted the improvements of Murray, the payments ceased. The Leeds firm forged ahead of all competitors, and it was largely because of Murray's introduction of " sponge weights " to improve the fineness of spinning, and his invention of " wet spinning." They revolutionised the business.

Some writers have told a " sob story " about Murray tramping from Stockton to Leeds, but there was plenty of wheeled and pack-horse traffic on the Great North Road at that time, and a likeable and adaptable traveller such as Murray, would have no difficulty in getting " lifts."

The story about the landlord of the Bay Horse Inn, Chapeltown, giving him a night's lodging without payment may be true, but he was taking no risk, because Murray would have his " kit of tools." Craftsmen had to carry their tools about in those days, for tools were then very " hard to come by." The Inn is about half-an-hour's walk from the mill.

It was customary for millwrights and others skilled in the mechanical arts to travel up and down the country and, as everything had then to be done by hand, such skilled knowledge was an asset of greater value than is a good banking account to-day.

Murray's first introduction to an engine, which experience some may think qualified him for the title " Engineer," came in 1791, when he took charge of a Savery type pumping engine at Marshall & Benyon's flax mill. Part of the building is still standing in Marshall Street.

This engine pumped water from Benyon's Beck, now covered up, to an elevated tank from which it fell on to the buckets of an overshot waterwheel. The arrangement may seem roundabout, but it gave that regular even turning of the spindles which was necessary for fine spinning of flax. Up to that time the location of all power-driven mills had been fixed by natural water power.

Murray not only studied his own engine, but he could see another of Newcomen's which Emmett had installed at Rothwell about 1776, also later on he had opportunities of seeing a rotative engine that Boulton & Watt installed at Marshall's.

It is of interest to note that the first experiences which James Watt and Matthew Murray had with engines was with those of the Newcomen and Savery types. As the Savery was the older of the two, this appears to indicate that there was a Savery engine in Leeds before the Newcomen engine was installed at Rothwell.

It must be remembered that John Smeaton was a Leeds Engineer and that he had improved steam engines by fifty per cent. before Watt began to make engines. Murray was therefore on " fertile ground."

When Murray first started in business with David Wood, in 1795, their workshop was in Mill Green, then a centre of activity because water-power was available from a " goit " of Holbeck Beck, and there were mills for cloth, flax and flour as well as malthouses, etc. It is possible that one of the very old buildings that still exist in Mill Green housed Murray's firm.

They carried out mill gearing, millwright work, and made textile machines, and it was there and in Marshall's mill that the business of making flax machinery originated, which later was developed by Fairbairns and Lawsons.

They afterwards took as partners James Fenton, a financier whose family had owned the coal mines at Rothwell, and William Lister, a millwright, of Bramley.

Murray was the driving force of the firm, and his skill in design and boldness in entering new fields of manufacture caused them to start making engines. In 1796 the firm moved to a large site in Water Lane, and Murray built an engine to drive the works.

In 1802 Murray built a new erecting shop like an engine cylinder, with entrance like a valve chest, and by doing so, gave a sidelight to his character. Such an unusual shape must have involved the firm in greater expense than a rectangular building, but he had his way, and the " gesture " is interesting, and it also shows his " force of character." All creative engineers are artists, although they may not know it.

Murray's artistic temperament was shown in the design of the beams and the framing, etc., of his engines, which John Farey and other contemporaries describe as " elegant." John Farey was the engineer who wrote authoritatively at that period, and he knew Murray well, because he was for a time in charge of Marshall's mill.

The erection of the round building caused the works to be nicknamed " Round Foundry," and that name has " stuck " ever since in a way characteristic of Holbeck. The real purpose of the building was for the " fitting up " of engines, the foundries for " green sand " and " dry sand " castings being ordinary rectangular buildings.

" The Leeds Intelligencer " of 11th July, 1796, says :—" Murray & Wood desire to inform their friends and the public in general that they have erected and opened a foundry in Water Lane."

The round building was burnt down in 1872, and most unfortunately along with it the drawings, correspondence, patterns, etc., of the Murray regime. A woman, still living in what was the old " Chequers Inn," which adjoined the works, saw it burning, and her " young man " rushed into the building just in time to save his kit of tools.

James Watt, Junr., in a letter of the 12th of June, 1802, mentions the " Rotundo " as being 100 feet diameter with a magnificent entrance. His use of that word is of passing interest, as a reminder that he had lived in France and was, in fact, a figure in the revolution until denounced by Robespierre just before the " reign of terror."

The purpose of his stay in Leeds was to spy upon the Holbeck firm, because they had become active competitors of Boulton & Watt. This went so far as sending engines to places near Birmingham.

Two years earlier William Murdock, the manager of the Soho works, and Abraham Storey, in charge of the foundry, had been hospitably entertained by Murray, who showed them round his works, and gave Murdock a sample of his own smithwork. They took particulars of his cylinder boring mills, which were very novel in having automatic screw feed, and bored very accurate surfaces. We know now from M. R. Boulton's letters that they were copied at Boulton & Watt's works.

It was because Murray used machine tools extensively that he was able to make such excellent engines, and his reputation for making tools grew until it became a special branch of manufacture. He exported them to France, Sweden, etc., and this was the beginning of the machine tool business which afterwards spread and made the engineering industry the accurate business it is to-day.

James Watt, Junr., tried to purchase a malthouse for the reason, as he explained to Boulton, that it—

"would enable us to overlook their whole yard, and, holding it, we might dictate our own terms."

The malthouse was owned by a widow who evidently found out why they wanted it. She would not sell, so they bought land adjoining the works. Watt wrote that they :—

"seem eligable speculations independent of the *ignoble* motives which dictate their purchase."

I wonder if the Leeds Corporation would have allowed a statue to James Watt to be erected in City Square if these letters had been known? It is only right to give as much prominence to a Memorial to Murray who did important constructive work for Leeds, as is given to the father of the man who tried to " scotch " Murray.

The land they bought, viz., Camp Field, remained vacant for very many years, and was eventually taken over by the Leeds Industrial Co-operative Society. It is bounded by David Street, which might well be re-named David Wood Street. I suggest that new street passing Victoria Square should be " Murrayway " or " Murrayhill." The euphony of the last is a reminder of the old name " Merrie Boys Hill."

Murray appears to have known the inventor of the hydraulic press, namely, Bramah, who was probably a relative of a millwright of that name then living in Churwell, near Holbeck. We do know that Murray was one of the first in the country to make hydraulic presses, and he probably supplied the press to Gott's Mill, which that firm was not able to use for some years because of opposition by the workmen.

Murray was undoubtedly the first to make a large hydraulic testing or " proving " machine, and in these days when one hears so much about the importance of scientific research and of testing materials, etc., it is interesting to know that he was engaged in such work over a century ago. This came about, as a result of General Wilson giving the firm an order for a testing machine to prove (*i.e.*, test) chain cables for the Navy. The machine was ordered to test up to 400 tons, but Murray built it for 1,000 tons, because, as he says in a letter to Simon Goodrich :—

" After it is made I intend to try a set of experiments with it, upon cast and wrought iron—to determine and compare the strength in different positions."

This machine was the forerunner of thousands of hydraulic testing machines made in Leeds by Greenwood & Batley and Joshua Buckton & Co. Testing machines associated with the name of Wicksteed, are to be found in nearly all the Technical Institutions and Testing Laboratories of the world.

From the point of view of the general public the most striking part of Murray's work was developing the steam locomotive, but unfortunately many writers will persist in making mistakes about it. They give John Blenkinsop the credit, when his work was entirely confined to a rack-rail ; it was Murray who designed and made the locomotives.

The order for these locomotives was given to Fenton, Murray & Wood by Charles Brandling, M.P., proprietor of Middleton Colliery, and he probably gave the order in 1810. Murray had first to expermient in the yard of the works in Water Lane. He had a locomotive ready by the following year.

This locomotive differed from any that preceded it by having two cylinders working on to two cranks set at right angles, and he adopted this arrangement in order to obtain a steady tractive effort and thus be able to start hauling the full load, regardless of what position the loco- motive might be in.

If Murray had used only one cylinder and one crank, as in the case of Trevithick's engines, then the locomotive could not have started the load from any position even if he had also employed a fly-wheel. A fly- wheel could only overcome the " dead point " difficulty when actually running at a good speed. There was no clutch mechanism as in motor- cars of to-day.

Although Murray's locomotive only weighed 5 tons, it could haul 90 tons, and this was because it had a toothed wheel which meshed into cogs on the Blenkinsop rack-rail. Without such cogs the engine could only have hauled about 15 tons. John Blenkinsop deserves great credit for being the pioneer of all rack-railways.

It has been said that Murray did not know that locomotives could haul loads with smooth wheels on smooth rails, but that assertion is wrong, because it was known to have been done by Trevithick.

Murray saw Trevithick when in London in 1809, when he went to receive the gold medal at the Society of Arts for his flax hackling machine. They had business relations together until Trevithick went to Peru in 1816.

It must be remembered that Murray and Blenkinsop were pioneering when cast iron was practically the only available metal for the rails, wrought iron being relatively expensive. To reduce breakages the loco- motive had to be of light weight, namely, 5 tons, and to haul 90 tons would have required five or six journeys with plain rails.

In view of what happened during the late war in rapid development of inventions, it is interesting to note that it was similar abnormal conditions that brought Murray's locomotives into regular commercial use in 1812. The long conflict with Napoleon caused inventive creative work of engineers to be in demand, and it was this fact which largely brought about the industrial revolution. The late war has brought about another industrial revolution in our time.

All food stuffs, including, of course, food for horses, went up to very high prices, and as Charles Brandling was under obligation by Act of Parliament to deliver annually a certain amount of coal to Leeds, he naturally looked about for another way of hauling his waggons of coal. Neither he nor his manager, John Blenkinsop, then called a " Viewer," knew much about engines, and they naturally went to the recognised authority of the district, namely, Matthew Murray.

Having talked the matter over with him, Blenkinsop, with the assistance of a certain John Straker, took out his patent for a rack-rail. They laid the cast-iron rails on an old and well consolidated waggon-way that was first built in 1758 to connect Casson Place, near Leeds Bridge, to the colliery " Day hole " at Belle Isle, below Middleton.

The total weight of the cast-iron used was about 500 tons, and the castings were made in the foundry on Carr Moor Side, Hunslet, close to the line. This foundry is the oldest in Leeds, and one of the oldest in the North of England, and for many years it had been occupied in supplying castings to Middleton and neighbouring collieries. It was then owned by John Goddard and a son of Titus Salt, the grandfather of Sir Titus Salt, builder, of Saltaire.

The cast-iron rails were taken up about 1860, and sold as scrap to my grandfather, Richard Kilburn, who had the same foundry, and his son, the late Richard Kilburn, of Roundhay, melted them down in the same " air furnace " from which the original metal had been poured half a century before. About 35 years ago, at the time of the Chicago Exposition, American engineers were very anxious to obtain a piece of the old rack-rail, and Commissioners were at the foundry for several days, raking over old metal, etc.

Most of the rails were made 3 feet long, but some were 6 feet, and one of the latter weighing about 170 pounds is now in the Leeds Museum, to which it was presented by Miss Maud, of Middleton Lodge, in which house Charles Brandling, M.P., lived about 120 years ago. This house is unique in having a full-sized " cock fighting pit " in the entrance hall.

Another house of interest is that which Murray built in 1802, which still stands, in a triangle of railway lines, and can be seen from the main Midland line just before running into Leeds. The title of the house is " Holbeck Lodge," but with their usual facility for applying nicknames, Holbeck folk of a century ago called it " Steam Hall," because Murray heated it with steam-pipes.

Being the first house to have such a form of central heating, it has an added historic interest, especially to Americans and Canadians, whose houses to-day are usually heated by a central system. After the Murrays left the house, my grandfather, Richard Kilburn, had it, and afterwards, when he built Malvern House, Beeston Hill (now the Vicarage), about 70 years ago, he installed central heating there also.

In 1812 Luddite rioters came to Holbeck Lodge to interview Matthew Murray with a view to forcing him to stop making labour saving machinery. He was away at the time, but Mrs. Murray—a masculine type of woman—fired a pistol from a bedroom window, and the men dispersed without doing any damage to the house or works.

Finally it is appropriate to say a word about the obelisk in Holbeck Churchyard, which, being of cast-iron and painted drab colour, is certainly not handsome. It may be remarked, however, that Murray did his creative or constructive work in the " age of cast-iron," which preceded that of wrought iron, and therefore the material of the obelisk is in that sense appropriate.

It was a " labour of love " on the part of his workmen, for they made the patterns and castings, and did the machining in the " Round Foundry." Murray was a typical example of the old patriarchal type of employer, who called his men by their Christian names, and long after he died it was an asset for any mechanic, millwright, pattern maker or moulder, etc., to be able to say that he had worked with " t'old Mattha Mirror," as he was called.

By the helpfulness in imparting information he was a Father to those who worked with him, and he was also " The Father of Leeds Engineering." Furthermore, I am convinced that in future he will be recognised as one of the great " Pioneers of Mechanical Engineering."

This gives me the cue to say that Murray's success was partly due to the able co-operation of men he gathered round him, for he had the faculty of a born leader in being able to choose clever assistants. Benjamin Hick and Samuel Owen, who made engineering history in Lancashire and Sweden, are two engineers trained by him.

It must be remembered that before Murray came to Leeds, the district had for generations been a centre for millwrights, smiths and others engaged in work connected with the weaving and fulling of cloth ; the mining of coal and ironstone ; the grinding of flour, etc.

There were craftsmen skilled in mechanical work, and Murray and other leaders had therefore much traditional skill to draw upon. The whole West Riding has always been a breeding ground, if one may use the term, of people with creative constructive ability. A Scottish professor of engineering has stated that West Riding mechanics are the best in the world.

Unfortunately a social system has developed, in which, chaffering in shops and routine work in offices, etc., receive relatively better financial returns than the exercise of creative constructive work. A voluble salesman is better considered than, say, an inventive engineering draughtsman. If Leeds were to discover to-day another engineer of the type that Murray was in his day he would be an asset of great price.

EXHIBITS AT THE LEEDS TOWN HALL.

The following exhibits were shown in Leeds during Tercentenary Week, 1926, and the list is given as a possible guide for other exhibitions :—

Portrait of Matthew Murray in colours, and copied by Mr. E. Kilburn Scott from the original now in the possession of Mr. F. J. March.

Model of Murray's flax hackling machine, awarded a gold medal by the Society of Arts in 1809. Lent by Mr. R. Murray Thomson.

Murray's Gold Medal. Lent by Mr. F. J. March, Oakbrooke, Derby.

Journals of the Society of Arts of 1809, with correspondence and a description of the hackling machine.

Picture of the Coal Staithe in Gt. Wilson Street with Murray's locomotive hauling and pushing wagons. Lent by Mr. Rhodes Calvert.

Photograph of the best model of Murray's locomotive as in the Science Museum, South Kensington.

Model of Murray's locomotive and piece of rack rail 3 ft. long, used on Blenkinsop's railway. Lent by Mr. T. Harding Churton.

Reprints of Acts of Parliament of 1758, 1759, 1793, and 1803, giving powers for a wagon road over Hunslet Moor from the Middleton Colliery. Lent by Mr. Emsley, solicitor, Leeds.

Patent No. 1752 of 1790 for spinning and drawing frames adopted at Marshall & Benyon's Mill, Holbeck.

Patent No. 1971 of 1793 for preparing and spinning flax, hemp, tow, wool and silk. A carding engine is also described.

Patent No. 2327 of 1799 and No. 2632 of 1802, describes details of steam engines and boilers.

Patent No. 2632 of 1803 shows a box or three-ported slide valve for controlling the supply and exhaust of steam to an engine cylinder.

Indenture of agreement dated 1772 for the sale of Hunslet Foundry of Low Moorside, one of the principals being Titus Salt, the grandfather of Sir Titus Salt. Blenkinsop's rails were cast and melted down there. Lent by Mr. Matt. Carr, Roundhay.

Journal of the Society of Arts of 1876, pp. 443 and 943, with account of the steam boat fitted with high-pressure engine, boiler and paddle wheels, by Fenton, Murray & Wood. Lent by Mr. E. Kilburn Scott.

Letters of James Watt, Junr., dated 12th and 14th of June, 1802. Mention is made of the round building, and Watt urges the purchase of land and buildings adjacent to the premises. From Watt Collection.

Photograph of Thomas Tannett, partner in the firm of Smith, Beacock and Tannett. Lent by Mr. T. Tannett, solicitor, Leeds.

Photograph of Robert Beacock, of Smith, Beacock & Tannett. Lent by Mr. W. Beacock Croysdale, Harrogate.

Photographs of Alfred Krupp, founder of the Essen Works, and J. A. Krupp, his son. Lent by Mr. Gaskell Walker.

Model of Murray's " straight line motion " engine and line drawing of Murray's boring machine at Chaillot, France, dated 1822. Lent by The Science Museum.

THE STORY OF HUNSLET FOUNDRY.

E. Kilburn Scott, A.M.Inst.C.E.

From " The Foundry Trade Journal " of July 29, 1926.

The artificers who helped to build Kirkstall Abbey in the twelfth century would be the first to work iron in Leeds, and although it is possible they did so at Kirkstall with charcoal as fuel, it is more than likely that they also employed the coal which outcropped at Beeston and Middleton, for there was a bridle path from the Abbey to Beeston.

When the Knights Templars were at Temple-Newsum they would need artificers accustomed to working iron to take care of their armour, etc., and Foundry Lane, which is near, is a present-day reminder of the fact.

In 1669 coal was being worked at Middleton on a scale sufficient to justify the striking of special coins, because a halfpenny found in Bramley has one one side " Francis Conyers of Middleton in Yorkshire, 1669," and on the other " for use of ye coal pits."

In the " Annals of Yorkshire " there is an account of a murder in 1678 of a *colliery owner* named Leonard Leurr, who had filled the office of minister of Beeston Chapel during the time of Cromwell, and was therefore a prominent man.

These refeences are given in order to show that a traditional skill in coal mining and iron working was growing up in Leeds, and especially South of the River Aire, long before the commencement of the industrial era at the beginning of the nineteenth century.

An interesting document of that period is the Polling List of the Parliamentary Election of 1807, a very critical year in English history, because of the Napoleonic wars. In it there is a record of the responsible burgesses who went to York to vote, and it gives their addresses and trades or professions.

The Hunslet polling list has the names of two millwrights, namely, John Jubb and David Trickett, and three ironfounders, John and William Gothard, and — Salt, showing that the various people who made engineering history in Hunslet had not yet come on the scene. As a matter of fact, most of them date from after the *middle* of the last centuyr.

In the Holbeck list there is Matthew " Mirror " (*i.e.*, Murray), described as an engineer, and the only one in the Leeds district to be so described, also his partners, James Fenton and David Wood, who are called ironfounders. In the Leeds polling list there is also an ironfounder called Lister, who was for a time a sleeping partner of Fenton, Murray and Wood. His name appeared on a document shown in the Matthew Murray Centenary Exhibit at the South Kensington Museum.

Matthew Murray was really the " Father of Leeds engineering," because it was he who first originated in Leeds the design and manufacture of spinning machinery, stationary steam engines, locomotives and of machine tools. The works of his firm, Fenton, Murray & Wood, in Water Lane, Leeds, were built about 1800 and had two foundries—one for green sand and one for dry sand moulding, with air furnaces to melt the iron, also a small cupola.

There were several other foundries in Leeds, and probably the oldest was that on Hunslet Moor, which the present owners, Denison & Sons, Limited, weighting machine makers, have occupied since 1899, when they took it over from R. Kilburn & Sons.

The first particulars of Hunslet Foundry, of which there is documentary record are in an agreement dated August 27, 1772, which relates that the executors of Robert Howitt sell to Timothy Gothard, ironfounder, of Hunslet, and Titus Salt, cast-steel maker, of Sheffield, the foundry and hereditament and tenements, etc.

Assuming that the foundry was in existence for a period of, say, thirty years before 1772, then it has had a continuous existence of nearly 200 years, which probably makes it the oldest in the Leeds district, if not one of the oldest in Yorkshire.

The agreement, written on parchment, measures 3 ft. square, and is the property of Mr. Matthew Carr, a great-great-grandson of the Gothard mentioned therein. For over a century it was kept in an old Hall, on Carr Moorside, Hunslet, a Tudor building having mullioned windows and coats-of-arms over its large open fireplace.

Thomes Fenton, an ancestor of James Fenton, the financial partner of Matthew Murray, is believed to have lived there when he was mining coal at Middleton and Rothwell. In Batty's " History of Rothwell " it is mentioned that one of the first pumping engines in the country was put down by Thomas Fenton in 1750. It was one of Emmett's inventions and worked on the atmospheric principle. This was before the time of Boulton and Watt.

The agreement is witnessed by Jas. Newport and T. Newport, and some of the wording reads quaintly :—

" On the day and year first within written, full possession as well as of the Tenements & Hereditaments as also of the Goods & Chattells herein mentioned to be hereby bargained and sold, were given to the said Titus Salt, and the delivery of the key of the outer door of the said Tenements and of one hammer part of the said Goods & Chattells in the name of possession and seizen."

The following are facsimile signatures on the document :—

The Timothy Gothard who signed the agreement died in 1805, at the age of 82 years, and his share in the foundry descended to his son John, who purchased the other share from Daniel Salt, the son of Titus Salt, the signatory to the agreement.

This Daniel Salt was a drysalter and ironmonger (which in those days meant also iron merchant) in Morley, but he moved to Bradford and became a wool stapler and worsted manufacturer. It was his son Titus, born in Morley, September 20, 1803, who first made alpaca, founded Saltaire, and became Sir Titus Salt.

John Gothard died in 1824, and then his son-in-law, John Gledhill, carried on the foundry until his death in 1855. He had three sons who, by a singular coincidence, died in three succeeding years. After the father's death the property came to the three daughters, the eldest of whom married an Excise officer, and she was the mother of Mr. Matthew Carr.

As the daughters could not carry on the business they sold it in 1855 to the writer's grandfather, Richard Kilburn, machine maker, of Holbeck, who had for some years been obtaining castings from the foundry. At the time his eldest son, Richard, was 14 years of age, and at school, but he was forthwith put into the foundry, and he managed it very successfully for nearly half a century.

The old foundry was famous for a special quality of iron almost like a semi-steel, which was used by many engineering firms for the high-pressure cylinders of hydraulic presses. This was partly made from old cannon taken from dismantled forts, etc.

The metal was melted in a specially constructed air furnace, run by an old-time character called Dave Hartley, who used to light it up about 3 o'clock in the morning. When he died the furnace stopped, because no one could be depended on to start work at that hour.

The first Blenkinsop rack rails which were laid down on the old waggon way (built by C. Brandling in 1758) from Middleton to Hunslet, were cast in this foundry in 1811, and when they were taken up in 1862 the late Richard Kilburn used them as scrap. Copies of the rack rails are in South Kensington Museum. The 3½ miles of line, required about 500 tons of metal, and as there would be many breakages it was convenient to make them in a foundry that adjoined the line.

John Blenkinsop, the patentee of the rack railway, appears to have designed the first rail. One that was preserved for about a century at Middleton is now in the Leeds Museum. It is 6 ft. long, and cruder in design than the 3-ft. length of rack rail. *See* page 71.

Rails of wrought iron first began to be used about 1820, and before then the problem was to make rails of cast iron strong enough to withstand the pounding weight of a locomotive and yet not cost too much. As the success of the Matthew Murray's locomotives depended on the strength of the cast-iron rails he probably had a say in the final design.

NOTE.—When the Stockton and Darlington Railway was first laid down in 1825 the price of cast iron rails was £5 10s. 0d. a ton and malleable iron rails about £12. The directors therefore instructed G. Stephenson to order 800 tons of each, although he wished to use only malleable. Robert Stevenson, of Edinburgh (grandfather of R.L.S.), reported in 1818 on wrought iron rails used at a colliery near Carlisle, and on 22nd Dec., 1820, Birkinshaw, of Newcastle-on-Tyne, took out a patent for the prototype of the bull-headed rail.

1765. Matthew Murray, born at Newcastle-on-Tyne, had an ordinary school education.

1779. Apprenticed to a smith, who probably did general millwrighting and mechanic's work.

1785. Married Mary Thompson, of Whickham, County Durham, a tall woman of strong character.

1786. Worked as journeyman mechanic at Stockton-on-Tees, where he had experience with machines for a flax mill at Darlington.

1788. Trade being slack, went to Leeds, and put up at the Bay Horse Inn, Chapeltown, and obtained work with John Marshall at Scotland Mill, Adel. Suggested improvements to machinery and received a present of £20. Sent for Mrs. Murray and lived in cottage on Blackmoor.

1790. Took out patent No. 1752 for machine for spinning yarn from silk, cotton, hemp, tow, flax and wool. His machines were installed in a new mill of Marshall & Benyon in Holbeck. This was driven by a water wheel supplied with water from the Beck by a Savery engine. Murray had charge of this engine and was thus able to study its working.

1793. Took out patent No. 1971 for apparatus and machines for preparing and spinning flax, hemp, tow, wool and silk. The specification describes a carding engine. Murray's introduction of " sponge weights " and the " wet spinning " of flax revolutionised linen manufacture and enabled John Marshall's firm to assume the lead in the industry.

1795. Became a partner with David Wood, and later they took James Fenton. Murray designed and erected engines ; Wood directed the works, and Fenton the office. For a time they had a sleeping partner called William Lister. The first workshop was in Mill Green, Holbeck, but soon removed to larger premises in Water Lane. Boulton & Watt's Soho Foundry was opened January, 1796.

1799. Took out patent No. 2327, and this showed a method of self-regulating the draught and steam pressure of a boiler. He received very hospitably William Murdock and Abraham Story, of Boulton & Watt, Soho Foundry, Birmingham. They were shown the shops and foundry of Murray's works, and took particulars of boring mills and other machines. Murdock was given a sample of Murray's own smithwork.

1800. Murray and his wife called on Murdock at 13, Foundry Road, Soho. The firm refused him admission to the Soho Foundry, although this had been promised when Murdock and Story visited the works in Holbeck.

1801. Took out patent No. 2531 for constructing the air pump and other parts belonging to steam engines so as to increase power and save fuel. This shows a form of mechanical stoker. The patent was opposed by Boulton & Watt. Details of his parallel motion were communicated by Murray to James White, who published the idea in *New Century of Invention*. The latter thus received credit for having invented it.

1802. Took out patent No. 2632, for improved steam engines to produce circular powers and machinery belonging thereto applicable for drawing coals and other minerals from mines, for spinning cotton, flax and wool, or for any other purpose requiring circular power. This patent shows one of Murray's important improvements, namely, the box or three-ported slide valve for steam engines. To accurately machine the parallel surfaces of these valves he made a planing machine which is believed to have been the first.

A large round building was erected, which caused the works to be called the " Round Foundry," and this building, is mentioned by James Watt, Junr., who was in Leeds, inquiring into the activities of Murray's firm.

1803. Boulton & Watt bought land in Camp Field next to the works of Fenton, Murray & Wood, in order to handicap them. See various letters written by James Watt Junr., to M. R. Boulton, and now in the Boulton & Watt Collection, Birmingham.

1804. Murray built a house, and centrally heated it with steam ; from which fact it received the name " Steam Hall." Commenced correspondence with Simon Goodrich, engineer and mechanist to the Navy Board. Fenton, Murray & Wood carried out many contracts for the Navy.

1808. Richard Trevithick's single cylinder locomotive ran on a circular track at Euston Square, London. This would be seen by Charles Brandling, M.P., of Middleton Colliery, Leeds, when in London in connection with his Parliamentary duties.

1809. Murray went to receive the gold medal given by the Society of Arts for his machine for hackling flax. He probably met Trevithick whilst in London, as they afterwards did business together.

1810. Murray commenced experimenting with steam traction at his works. He had a locomotive ready in the following year, for there is a representation of it on Leeds pottery that is dated 1811.

1811. John Blenkinsop took out patent No. 3431 for a rack railway, and he laid cast iron rails, each 3 ft. long and 4 ft. 1½ in. gauge, on the wagon way between Middleton and Hunslet. These rails were cast in Hunslet Foundry and afterwards melted down there.

1812. The first public demonstration of Murray's locomotive on August 12th. It weighed 5 tons and hauled 90 tons, and cost £380, including a Royalty of £30 paid to Trevithick. The names of the first two locomotives were " Salamanca " and " Prince Regent," the latter name because of the Regent's birthday. One remained in use until 1835, and it was afterwards exhibited in a shed at Belle Isle, Middleton, until about 1860.

1813. The "Lord Wellington" locomotive was delivered on August 4th, and the "Marquis Wellington" on November 23rd, and George Stephenson inspected the locomotives on September 23rd. Two more locomotives were made for a rack railway connecting Kenton and Coxlodge Collieries to Walker-on-Tyne. George Stephenson afterwards built one that was practically a copy.

In this year Murray fitted a steam engine, boiler and paddle wheels to "The Experiment," an old French privateer, 50 ft. long, owned by John Wright. The boat plied for some years between Norwich and Yarmouth, and carried 107 passengers. The engine and boiler were later transferred to another boat which made a trip to the Medway, and this was probably the first sea trip made by any regular passenger boat.

Fenton, Murray & Wood were busy in the machine tool business, and they supplied cylinder boring machines to Portsmouth Dockyard and to engineering works in France.

1814. At this time Murray was giving much attention to the application of hydraulic pressure for various purposes, and he took out patent No. 3792, for a hydraulic press in which the ram and the top table moved in unison. He devised an instrument for indicating the pressure, also he made cable proving machines.

1815. A cross sectional working drawing of Murray's locomotive appeared in "*Bulletin de la Societe d'Encouragement pour l'Industrié Nationale,*" and from this the model in the Science Museum, South Kensington, was constructed in 1910. In a letter to Simon Goodrich, Murray suggested the use of paddle steamboats for the Navy.

1816. The Grand Duke Nicholas of Russia visited the "Round Foundry," and inspected the locomotives on the Hunslet to Middleton Railway, and he gave Murray a diamond ring. Murray's firm and the Leeds Pottery, etc., did a considerable trade with Russia and Sweden. He received presents from the Czar and the King.

1819. Murray lighted his own works and several streets of Leeds with coal gas and introduced improvements in the construction of gas retorts, condensers, etc.

1820. Thomas Gray, of Leeds, published "Observations on a General Iron Railway," advocating the building of rack railways, and he gave a good illustration of Murray's locomotive.

1822. A model of the locomotive taken to the Czar of Russia by Murray's only son, Matthew, who married there and died in Moscow. Another model was owned by John Blenkinsop, and passed to the Embleton family.

1824. In a letter to Goodrich, Murray mentions a hydraulic machine for testing chain cables which will work up to 1,000 tons, also that he is going to make tests on strength of materials.

1826. Murray died February 20th, and was given a public funeral. He was buried in the family vault in Holbeck Churchyard, which is marked by a plain cast-iron obelisk, made by his workmen.

INDENTURE OF APPRENTICESHIP.

The signatures given below are photographed from an indenture apprenticeship agreement of Fenton, Murray & Wood dated 1806, which belonged to John Kendell, of Holbeck, who was apprenticed with the firm for seven years. He became a " journeyman " soon after Murray's locomotives started running between Hunslet and Middleton, and probably assisted in their construction.

Two of his sons worked at the " Round Foundry," and one named Thomas came to London after 1843 when the firm of Fenton, Murray & Jackson closed down. After being a fitter with the Gas Light & Coke Co., at the Haggerston gas plant for several years, he went to work for a relative, namely, Joseph Boulton, who had an engineering works in French Place, Shoreditch.

The business eventually passed to Thomas Kendell and then to the son, Thomas Boulton Kendall, and the sons of the latter now carry on the business. It is the oldest printers' engineers firm in London and well-known for " guillotine " and other cutting machines.

Mr. Thomas Boulton Kendall was the first in London to make a band knife machine for cutting layers of cloth, and it was for a Frenchman called Aublet, who had a business in Curtain Road, London. The band knife ran at 1,800 feet per minute over three wheels each two feet diameter. Later machines had five wheels so as to make the path of the knife more circular.

It is interesting to note that the two first band knife machines in this country were made by Greenwood & Batley, of Leeds, and by T. B. Kendall, of London, a descendant of a Leeds engineer. It was the introduction of the band knife that helped John Barran to establish the wholesale clothing trade for which Leeds is famous.

The indenture is now owned by T. Boulton Kendell, of Ilford, London.

BIBLIOGRAPHY.

"Leeds Mercury," July 20th, 1803. Advertisement in which Matthew Murray challenges Boulton & Watt to comparative tests of their respective engines.

"Repertory of Arts," 1803, second series III., p. 238. Mentions that Murray's patent of 1801 was set aside by *Scire facias* at the instance of Boulton & Watt, on the ground that a portion was known to them.

"Journal of Natural Philosophy," by Nicholson, 1805, XI., p. 93, contains illustration of Murray's inverted beam engine made in 1805.

"Essay on the Teeth of Wheels," 1808, by Robertson Buchanan, says : " The ingenious Mr. Murray, of Leeds," shows how to set out teeth of wheels.

"Leeds Mercury," June 27th, 1812. Report of the opening of the railway line with Blenkinsop rack rail from Hunslet to Middleton Colliery, and particulars of performance of Murray's locomotive.

"Leeds Mercury," July 18th, 1812. Description of the locomotive, with an illustration.

"Leeds Mercury," June 24th, 1813. Mentions the fitting of Murray's engine to a steam boat to ply between Yarmouth and Norwich. This was done in the Canal basin, Holbeck.

"The Costume of Yorkshire," 1814. An aquatint entitled " The Collier," by George Walker, shows in the background Murray's locomotive, carrying two men and hauling four wagons.

"Journal of Society of Arts," 1809, Vol. XXVII., pp. 148-153. Description of model of Murray's invention of a machine for hackling flax, for which he was awarded a gold medal by the Society. This model is now in the possession of Ralph Murray Thompson.

"Repertory of Arts," 2nd Series, 1815. Description of Murray's hydraulic press.

"Bulletin de la Societe d'Encouragement pour l'Industrie Nationale," 1815. Contains the best contemporary drawing of Murray's locomotive, and it is the one from which the model in the Science Museum was made.

"Bulletin de la Societe d'Encouragement pour l'Industrie Nationale," 1823. Gives a description and drawing of Murray's cylinder boring machine, supplied to the Chaillot Engine Works of M. Pêrier.

"The Encyclopædia," a Universal Dictionary, by A. D. D. Rees, F.R.S., 1819-20. Article on Steam Engines, written by John Farey, in 1815. Describes the rack rail locomotive at Newcastle-on-Tyne. There is also a description of the Works of Fenton, Murray & Wood.

"Encyclopædia Britannica," 6th edition, 1817. Drawing, by John Farey, of one of Murray's cylinder boring machines.

"London Journal," 1821. Murray's furnace for preventing smoke.

" OBSERVATIONS ON A GENERAL IRON RAILWAY," by Thomas Gray, 1821, gives a good drawing of Murray's locomotive. The rack rail of Blenkinsop is not correctly shown.

" TECHNICAL REPOSITORY," by Gill, 1822. Gives the earliest complete drawing of a cylinder boring machine made by Fenton, Murray & Wood. This was taken from a French publication.

" MECHANICS' DICTIONARY," by Grier, 1823. Particulars of Murray's patents.

" PRACTICAL TREATISE ON RAILROADS," by Nicholas Wood, 1825, p. 128. Description and drawing of Murray's locomotive with *enlarged* boiler and chimney. The teeth of the rack rail are shown correctly on Fig. 1 of Plate IV.

" THE STEAM ENGINE," by John Farey, 1827, p. 692, mentions that the D slide valve formed part of one of Murray's engines as early as 1810. Describes and illustrates an engine for driving a Bark Mill in Bermondsey, which was fitted with Murray's cycloidal straight line motion.

" THE STEAM ENGINE," by Thomas Tredgold, 1827, p. 37. Describes Murray's improvements on the steam engine. Page 217 describes and illustrates the D slide valve.

" HISTORY AND ANECDOTES OF THE STEAM ENGINE," by Robert Stuart, 1829, Vol. II., pp. 441-3. Particulars of Murray's improvements in steam engines, illustrated by three plates.

" HISTORY OF THE STEAM ENGINE," by Elijah Galloway, 1829. Gives particulars of Murray's engines supplied to St. Peter's Quay, Newcastle-on Tyne. These were fitted with D slide valves.

" MECHANICS' MAGAZINE," 1829, p. 131. Messrs. Ellis, Sander, Booth & Kennedy visited collieries in North of England in 1825, where locomotive steam engines were employed. They refer to the excellent operation of Murray's locomotives.

" REPORT TO THE DIRECTORS OF THE LIVERPOOL AND MANCHESTER RAILWAY ON THE COMPARATIVE MERITS OF LOCOMOTIVES & FIXED ENGINES AS A MOVING POWER," by J. Walker and J. V. Rastrick, 1829. They saw Murray's locomotives in November, 1828, and found that on an average, 2¾ lbs. of coal was consumed for each ton of coals conveyed one mile, at the speed of 4 miles per hour.

" MECHANICS' MAGAZINE," 30th October, 1830. Describes the engine supplied by Fenton, Murray & Wood to Francis B. Ogden, United States Consul at Liverpool. One was fitted into a boat on the Mississippi.

" DESCRIPTIVE HISTORY OF THE STEAM ENGINE," by Robert Stuart, 1831. Ranks Murray's improvements on the steam engine as next to Watt.

" LEEDS MERCURY," January 29th, 1831. Obituary notice of John Blenkinsop, who died January 22nd, aged 48 years.

" STEAM MACHINERY," by Thomas Tredgold, 1840. Describes Murray's engine fitted with a second motion shaft and the valves driven by cams on a " layshaft "

" TREATISE ON THE STEAM ENGINE," by John Bourne, pub. 1842, p. 192. Describes Murray's ingenious mechanism for preventing smoke from boiler furnaces.

" ENGINEER AND MECHANISTS' ASSISTANT," 1843. Describes Murray's encycloidal parallel motion.

" VIEW OF THE COAL TRADE OF THE NORTH OF ENGLAND," by Matthias Dunn. Mentions that Blenkinsop, " aided by the advice and good suggestions of the late John Straker," took out a patent for a rack railway.

" LIFE OF GEORGE STEPHENSON," by Samuel Smiles, 1857, pp. 82-3. Description of locomotives built for the Kenton & Coxlodge Collieries by Fenton, Murray & Wood, and first run on 2nd September, 1813. George Stephenson examined these locomotives in detail.

" INDUSTRIAL BIOGRAPHY," by Samuel Smiles, 1863, p. 260. The earliest complete biographical notice of Murray, written whilst Smiles was living in Holbeck, and thus able to talk personally with those who had known him. In it J. Ogdin March mentions the early planing machine in Murray's works.

" MILLS AND MILLWORK," by Sir Wm. Fairbairn, 1865, Vol. II., p. 197. Gives credit to Messrs. Marshall & Murray as being the first to introduce machinery for spinning of flax.

" LEEDS WORTHIES," by Rev. R. W. Taylor, 1866. Gives a good biographical notice with voluminous footnotes by J. Ogdin March, a son-in-law of Murray.

" THE ENGINEER," 1874. Article about Samuel Owen, engineer for Fenton, Murray & Wood, in Sweden.

" JOURNAL OF SOCIETY OF ARTS," Vol. XXV., 1876, pp. 445 & 943. Notes on the introduction of steam navigation on the River Yare in 1813. A French privateer called L'Actif was fitted with a steam engine, boiler and paddle wheels, made at works of Fenton, Murray & Wood. A sea trip to the Medway was made by the " Telegraph," fitted with the same engine and boiler.

" OLD YORKSHIRE," by Wm. Smith, 1882, Vol. III., pp. 257-264. Contains an article by W. S. Cameron, and an excellent steel engraving of Murray given by John Rhodes, J.P., of Potternewton House.

" ANNALS AND CHARACTERISTICS OF DARLINGTON," by Longstaffe. Information about the spinning of flax by machinery invented by John Kendrew who took out a patent in 1787. These machines were seen by John Marshall, of Leeds.

" DICTIONARY OF NATIONAL BIOGRAPHY," pub. 1894, pp. 389-9. Well written article by R. B. Prosser. He mentions the patent of 1801 being set aside by Scire facias.

" LOCOMOTIVE MAGAZINE," 1898, p. 83. Reference to John Chester Craven.

"CATALOGUE OF LAND TRANSPORT—RAILWAY LOCOMOTIVES," in the Science Museum, London, compiled by E. A. Forward, A.R.C.Sc., M.I.Mech.E., 1923, pp. 12-13. Gives dimensions of Murray's locomotive.

"CATALOGUE OF STATIONARY ENGINES," in the Science Museum, London. Compiled by H. W. Dickinson, M.I.Mech.E., 1925, pp. 44-5. Gives accurate description of drawings and models of Murray's engines.

"TIMOTHY HACKWORTH AND THE LOCOMOTIVE," by R. Young, 1923, Gives particulars of Murray's locomotives on rack railway between Coxlodge Colliery and Walker-on-Tyne.

"TRANSACTIONS OF THE NEWCOMEN SOCIETY," 1924, III., p. 9. "Simon Goodrich and His Work as an Engineer," by E. A. Forward, M.I.Mech.E. Gives extracts of letters of Goodrich by Murray.

"THE NORTH EASTERN RAILWAY," by W. W. Tomlinson, p. 118. Mentions that Murray declined " to supply " locomotives to Stockton and Darlington Railway until such time as the high pressure locomotive had become established.

"RAILWAY MAGAZINE," March, 1925. Particulars of Murray's locomotive and illustrations of it hauling and pushing wagons at the coal staithe, Great Wilson Street, Hunslet.

"CENTURY OF LOCOMOTIVE BUILDING BY ROBERT STEPHENSON & CO.," from 1823-1923, by J. G. H. Warren. General discussion about the relative work of pioneers of steam locomotion. He mentions that Trevithick went to Peru.

"YORKSHIRE POST," 2nd Feb., 1926. Report of lecture on Matthew Murray at Holbeck Church Schools, by E. Kilburn Scott.

"YORKSHIRE POST " and " YORKSHIRE OBSERVER," 22nd Feb., 1926. Descriptions of Centenary Memorial Service at Holbeck Church.

"LEEDS MERCURY," 23rd Feb., 1926. " How Leeds Started Steam Boats : Murray's Work as a Pioneer," by E. Kilburn Scott.

"YORKSHIRE EVENING POST," 25th Feb., 1926. " An Episode in the Life of Leeds First Engineer," by E. Kilburn Scott.

"YORKSHIRE POST," 25th Feb., 1926. " Letters of Matthew Murray : Light on his Engineering Secrets," by E. Kilburn Scott.

"THE LOCOMOTIVE MAGAZINE," Jan., 1926. Letter by J. G. H. Warren, mentions that a spring safety valve was used on Murray's first locomotives.

"PIONEER RAILWAY BUILDING IN LEEDS," *Yorkshire Post,* Leeds, Tercentenary Supplement, 8-17th July, 1926. A chatty description of the first practical locomotive and of the first railway lines in Leeds.

"TRANSACTIONS OF NEWCOMEN SOCIETY," 1926, p. 33. " Early History of the Cylinder Boring Machine," by E. A. Forward. Gives particulars of Murray's boring machines, with two line drawings.

"THE LOCOMOTIVE," 15th February, 1926. Matthew Murray and the locomotive. A good description of the work of Trevithick and Murray.

' MODEL ENGINEER AND LIGHT MACHINERY REVIEW," 1st April, **1926.**
The Matthew Murray Centenary Exhibition at the Science Museum. An
excellent description of the various exhibits.

" FOUNDRY TRADE JOURNAL," July 29, **1926.** Story of Hunslet
Foundry, by E. Kilburn Scott.

" JAMES WATT AND THE STEAM ENGINE MEMORIAL," Vol. prepared for
the Committee of the Watt Centenary Commemoration, **1927,** by H. W.
Dickinson & Rhys Jenkins. Contains references to the relations between
Boulton & Watt and Matthew Murray.

" TRANSACTIONS OF NEWCOMEN SOCIETY," **1928.** " Matthew Murray,
a Centenary Appreciation," by G. F. Tyas, with copies of Murray's letters
to Simon Goodrich and of letters by M. R. Boulton and James Watt,
Junr., regarding Murray.

Although the above list is long, it may not be exhaustive,
and if any reader should know other references the writer
would be glad to have them.

A University degree represents no particular attainment except a
certain capacity for memory and reading of the most limited kind, and is
no particular proof whatever of all that we should mean by education.
—Archbishop Lang.

Every director of large enterprises must choose between competing
mechanical devices, must watch the course of invention, must be in the
fore with improvements.—Prof. Taussig.

Managers and owners of industrial plants who are inclined to regard
the question of the efficiency of their plants as a private and personal
matter are, to be frank, dangerously limited in their outlook, and those
who fail to return a fair percentage of their earnings to the industry itself,
in the form of plant improvement, are certainly doing the industry and
their own pockets a great deal of harm.—A. J. T. Taylor.

There is something wrong with a state of affairs in which technicians
do not earn as much as dustmen and men on the streets pushing barrows.
It would have to be altered or young men would refuse to be trained as
technicians. The Fuel Conference was opened with a flourish by three
prominent men, all of whom were lawyers. Would lawyers think of
asking three engineers to open a conference held by themselves?—E.K.S.

Some business men are smart. They know " the tricks " of the
trade. They draw up a contract with a tricky clause in it that gives
them more than their fair share of the profits. They have learned how
to do things that may be legal but that are not right. All this I call
" Tricknique." It is a vastly different thing from Technique. There
is no trickery in Technique. Technique means knowledge and skill and
experience, for the purpose of giving the public a better service.
—Herbert Casson.

CONTEMPORARY ENGINEERING CHRONOLOGY.

Before the middle of the eighteenth century, two leading figures in steam engine history were Thomas Savery (b. 1650, d. 1715), of Cornwall, who first used coal as a practical means of performing mechanical work, and Thomas Newcomen (b. 1663, d. 1729), of Dartmouth. The latter had a separate vessel to raise the steam and injected cold water into the engine cylinder to make a quick vacuum.

Although the engines which they made were of crude design, they did very useful work in pumping water from mines, etc. In 1769 John Smeaton (b. 1724, d. 1792), of Leeds, estimated, that over a hundred such engines were at work in the North of England, and over fifty in Cornwall.

Smeaton was then the leading engineer in the country, his scientific attainments being such, that in 1853 he was elected a Fellow of the Royal Society. He was the first to apply scientific methods to steam engine problems, and in 1768 he cut down the cost of fuel in a Newcomen engine by about 50 per cent.

Even in those early days Leeds and the West Riding generally were becoming known for mechanical enterprise. It was partly because of the reputation of John Smeaton which caused many people to consult with and visit him. The ease with which coal and ironstone were mined and the rivers harnessed also helped industrial development in the West Riding.

The following are some leading dates in engineering history with special references to Leeds. Only Mayors with scientific or engineering knowledge are mentioned.

1758. Act authorising waggon way between Hunslet and Middleton.
1759. Smeaton awarded the gold medal of the Royal Society.
1760. Leeds Pottery started in Hunslet by Humble, Green & Co.
,,   Robert Kay invented " drop box " for changing shuttle of loom.
1763. James Watt repaired model of Newcomen engine.
1765. Matthew Murray born at Newcastle-on-Tyne.
1767. Joseph Priestley came to Mill Hill Chapel, Leeds.
,,   James Hargreaves invented " Spinning Jenny."
1768. Leeds Library started.
,,   James Watt patented separate condenser steam engine.
1769. Richard Arkwright invented drawing roller frame for cotton.
,,   Smeaton experimented with atmospheric engine at Leeds.
1770. Leeds and Liverpool Canal commenced.

1771. Richard Trevithick born at Illogan, Cornwall.
,, Henry Maudslay born at Woolwich.
,, Smeaton founded first Engineers' Club in London, known as the "Smeatonian."
1772. Titus Salt joined Timothy Goddard at Hunslet Foundry.
,, Smeaton erected improved atmospheric engine at Long Benton Colliery, Northumberland.
1774. Joseph Priestley discovered oxygen, calling it "dephlogisticated air," thus founding modern chemistry.
1775. White Cloth Hall built in the Calls, Leeds.
,, Richard Arkwright built mill at Cromford, Derbyshire.
,, John Wilkinson made machine for boring engine cylinders.
1779. Samuel Crompton invented spinning "mule."
,, Kirkstall Forge started by the Butler family.
1780. Patent for applying crank to engine to obtain rotatory motion.
1781. George Stephenson born at Wylam, Northumberland.
,, William Hartley joined Leeds Pottery, then Hartley, Green & Co.
,, Joseph Hornblower patented compound steam engine.
1782. Watt patented the double-acting beam engine with parallel motion.
1783. Henry Cort method of puddling and rolling iron, using coal.
,, Boulton & Watt erected pumping engine at Rothwell Haigh.
,, First Leeds Literary and Philosophical Society, founded.
,, John Blenkinsop, first inventor of a rack rail, born.
1785. Edmond Cartright invented power loom.
1786. Timothy Hackworth born at Shildon, Co. Durham.
,, Matthew Murray in mechanics' shop at Stockton-on-Tees.
1787. Smeaton patented machine to extract oil from seeds.
,, John Marshall started flax-spinning at Scotland Mill, Adel.
1788. Symington's first steam boat on Dalswinton Loch, Dumfriesshire.
,, Low Moor Iron Works started.
,, Watt applied rotative steam engine for driving mills, etc.
,, Matthew Murray started with John Marshall, flax spinner, at Adel.
1789. Bowling Iron Works started.
,, William Jessop used edge rails on waggon way at Loughborough.
,, Beginning of French Revolution.
1790. Murray took out patent for spinning flax, etc.
1791. Public lighting of Leeds streets by oil lamps.
,, Birmingham mob destroyed Dr. J. Priestley's books and papers.
,, Marshall and Benyon started flax mill in Holbeck.
1792. John Smeaton died at Austhorpe, Leeds.
,, William Murdoch lighted his house, at Redruth, with coal gas.
1793. Murray patented machinery for preparing and spinning flax, etc.
,, Newcastle Literary and Philosophical Society founded.
,, Sir Samuel Bentham described a wood planing machine in his patent.
,, Stone chairs used for rails on waggon ways of collieries.
1794. Maudslay made compound slide rest for lathe.
1795. Murray started business with David Wood at Mill Green, Holbeck.
,, Joseph Bramah invented the hydraulic press.

1796. Benjamin Outram built plate ways in Derbyshire.

,, Soho Foundry opened by Boulton & Watt.

,, Aloys Senefelder discovered lithographic printing.

1797. Fenton, Murray & Wood started works in Water Lane, Holbeck.

1798. Leeds has 6 machine makers, a millwright and an ironfounder.

,, Part of Boulton & Watt's works lighted by coal gas.

,, Maudslay left Bramah, to start in business for himself.

1799. Murray's patent for regulating draught and pressure of boilers.

,, John Gott, first to make cloth on factory system, in Leeds, elected
    Mayor. An early industrialist to be in public life.

,, William Murdock and Abraham Storey visit Murray's works.

,, Founding of the Royal Institution, London, by Count Romford.

1800. Murdock introduces eccentric for working valves of steam engines.

1801. Murray's patent includes for a smoke consuming device.

,, Murray communicated his cycloidal straight line motion to James
    White, who later received a medal from Emperor Napoleon.

,, Act authorising Surrey Iron Railway from Wandsworth to Croydon.

1802. James Watt, Jun., in Leeds to discover reason of Murray's
    superiority in engine building.

,, Trevithick patented high-pressure steam engine and boiler.

,, First Factory Act passed.

,, Murray's patent for improved D slide valve.

,, Round building erected for " fitting up " shop, by Fenton, Murray
    & Wood, from which came the name " Round Foundry."

,, First Australian wool came to England.

1803. Fourdrinier's introduction of continuous paper-making machine.

,, Boulton & Watt purchased land adjoining works of Fenton, Murray
    & Wood, to prevent their extending.

,, Robert Stephenson born at Willington Quay, Newcastle-on-Tyne.

,, Cast iron plates used on Middleton Colliery waggon way.

,, William Symington's steamboat on the Forth and Clyde Canal.

1804. Samuel Owen went to Sweden to represent Fenton, Murray & Wood.

,, Trevithick's locomotive ran experimentally on Pen-y-darran plateway

,, Dr. Joseph Priestley died at Northumberland, Pennsylvania.

,, H. Jacquard invented loom for making figured fabrics.

,, Surrey Iron Railway opened for public traffic.

,, Murray built a house and centrally heated it with steam.

,, Coalition of Gt. Britain, Russia, Austria & Sweden against France

1805. Battle of Trafalgar, October 21st.

,, Rectangular wrought iron rail introduced.

,, Humphry Davy produced the electric carbon arc light.

1807. Election of Lord Milton, as Whig member for West Riding.

,, William Savage investigated printing inks and wrote on subject.

,, Maudslay patented the " table " type of engine.

1808. Trevithick ran a locomotive on edge rails in London.

,, James Nasmyth born at Edinburgh.

,, Rope haulage used on inclines at collieries.

,, Murray received gold medal from Society of Arts for model of flax
    hackling machine.

1811. John Blenkinsop obtained patent for rack rail.
,, Trevithick built the first steam threshing machine.
1812. Henry Bell launched the " Comet " steamboat on the Clyde.
,, Gas Light and Coke Company, London, obtained charter.
,, Direct-loaded safety valve first used by Murray.
,, Feed pumps driven by motion of Murray's locomotive.
,, Murray's locomotives running regularly on Hunslet to Middleton railway. A 5 ton locomotive hauled 90 tons.
,, Luddite riots in Leeds.
,, Samuel Lawson and J. Walker started flax spinning.
1813. Murray supplied locomotives to Kenton and Coxlodge Collieries railway line.
,, George Stephenson came to Leeds to see Murray's locomotives.
,, Hedley built " Puffing Billy " locomotive, geared to track wheels.
,, Murray fitted engine and boiler into steam boat for Wright, a Quaker, of Yarmouth.
,, Murray supplied cylinder boring machine to Portsmouth Dockyard.
,, First National Schools opened in Leeds.
1814. George Stephenson's first locomotive at Killingworth Colliery.
,, Murray patented hydraulic baling press and pressure indicator.
,, Joseph Bramah died at Pimlico, London.
,, A metal planing machine used by Murray to plane his valves.
1815. Battle of Waterloo, June 18th.
,, Wheat and other food stuffs selling at very high prices.
1816. Grand Duke Nicholas of Russia inspected Murray's locomotives.
,, Richard Trevithick went to Peru.
,, Completion of the Leeds and Liverpool Canal.
,, Industrial depression in Great Britain.
1818. Philosophical and Literary Society founded in Leeds.
,, Institution of Civil Engineers founded.
,, The " Savannah " used steam engine for voyage across Atlantic.
1819. Leeds streets lighted by gas with plant made by Murray.
,, Birmingham Gas Light & Coke Co. started.
,, Act prohibiting employment of children under nine years of age.
,, Riot in St. Peter's Field, Manchester.
1820. Birkinshaw patented wrought iron rail of Tee shape.
,, Thomas Gray wrote " Observations on a General Iron Railway."
,, William Hey, F.R.S., Mayor of Leeds.
,, Demand for Parliamentary reform and Free Trade.
1821. General improvement in trade of West Riding.
1822. Fenton, Murray & Wood sent cylinder boring machine to Chaillot.
,, Model of Murray's locomotive taken by his son to Czar of Russia.
,, Riots at Shipley against use of machinery.
1824. Murray made hydraulic machine for testing chain cable to 1,000 tons.
,, J. P. Poucelet improved undershot water wheel.
,, Joseph Aspdin, of Leeds, granted patent for making Portland cement.
,, Trade revival ; many joint stock companies started.
1825. Stockton and Darlington Railway opened for passenger traffic.
,, Trades Unions (for fixing wages) made legal.

1825. William Sturgeon invented the electro-magnet.
  ,, John Marshall, of Holbeck, elected Member of Parliament.
  ,, Commercial crisis; sixty banks suspended.
1826. Murray died; funeral at Holbeck Church.
1827. Richard Trevithick returned from South America.
  ,, Act for establishing gasworks in Leeds.
  ,, Benoît Fourneyron invented radial outward flow water turbine.
  ,, Joseph Maudslay patented oscillating engine.
1828. James B. Neilson's hot blast process halved cost of pig iron.
  ,, Peter Fairbairn started Wellington Foundry, Leeds.
  ,, Joseph Henry, of America, improved on Sturgeon's electro-magnet.
1829. Henry Booth suggested fire-tube boiler for "Rocket" locomotive.
  ,, Locomotive competition won by Stephenson's "Rocket" at Rainhill.
1830. First sewing machine made at St. Etienne, by Thimmonier.
  ,, Liverpool and Manchester Railway opened.
  ,, Power looms in general use for making woollen cloth.
  ,, Syphon oil cup lubricators used on steam engine.
1830. Second French Revolution.
  ,, Whig Ministry, with Brougham as Lord Chancellor.
1831. British Association founded in York.
  ,, Michael Faraday discovered electro-magnetic induction.
  ,, Henry Maudslay died at Lambeth, London.
  ,, John Blenkinsop died in Leeds.
  ,, Of 100 locomotives built to date, 90 were British, 6 American,
        2 French and 2 German.
  ,, William Hey, F.R.S., Mayor of Leeds.
1832. Reform Bill carried. Birmingham, Leeds, Manchester and
        Sheffield received right to send members to Parliament.
1833. Greenwood & Batley started machine making, at Armley, Leeds.
  ,, Trevithick died at Dartford, Kent.
  ,, Abolition of slavery in British possessions.
  ,, Samuel Lawson & Wm. King Westley patented screw for flax
        drawing and roving frames.
1834. Opening of railway between Leeds, Marsh Lane Station, and Selby.
1835. Steam whistle first used on locomotives.
  ,, Joseph Locke introduced double-headed rail.
  ,, Middleton coal-pit down to 420 ft. level.
  ,, Leeds Municipal Corporation Parliamentary Act.
  ,, Matthew Murray, Jnr., died in Moscow.
  ,, P. Fairbairn & Co. making jute machinery for Dundee.
1837. Todd, Kitson & Laird began building locomotives in Hunslet.
  ,, William Fourness obtained patent for ventilation of coal mines.
  ,, W. F. Cooke and C. Wheatstone introduced electric telegraph.
  ,, Strike at the works of Fenton, Murray & Jackson.
1838. First crossing of Atlantic by steamboat "Sirius."
  ,, Forty flax mills in Leeds, 6,000 employees.
1839. William Murdock died at Birmingham.
  ,, Corporation of Leeds bought the "Soke" rights for grinding corn.
  ,, Kitson & Laird start Airedale Foundry.

1839. James Nasmyth designed steam hammer.
,,   Self-acting planing machine made by Joseph Whitworth.
,,   Trade depression in all industries.
1840. Benjamin Gott, cloth manufacturer, died in Leeds.
,,   Joseph Whitworth made extremely accurate plane surfaces.
,,   Penny postage introduced.
1841. Robert Stephenson patented long boiler and inside cylinders.
,,   Foundation of Chemical Society.
,,   Schneider made steam hammer at Creuzôt, after Nasmyth's design.
,,   Nicholas Joseph Jonval invented parallel flow water turbine.
1842. North Midland Railway started, with station in Hunslet Lane.
,,   Women and children prohibited from working underground.
,,   William F. Cooke suggested block system for working railways.
,,   Local Improvement Act passed, with 391 clauses.
,,   William Howe invented link motion reversing gear.
1843. Fenton, Murray & Jackson closed the " Round Foundry."
,,   Nasmyth made his own steam hammer.
1844. Farnley Iron Works started.
,,   Group of employees took over the " Round Foundry."
,,   Intensive Railway speculation led by George Hudson, of York.
1845. Fairbairn's and other firms busy on railway contracts.
1846. Railway between Leeds and Bradford opened.
,,   Charles Gascoigne Maclea, engineer, Mayor of Leeds.
,,   Repeal of the Corn Laws, and Free Trade policy started.
1848. Leeds, Dewsbury and Manchester Railway opened.
,,   Foundation of Institution of Mechanical Engineers .
,,   George Stephenson died at Chesterfield.
,,   First Submarine telegraphic cable, to France.
1849. James B. Francis brought out mixed flow water turbine.
1850. Shaping machines introduced by Nasmyth and Gaskell.
,,   Smith, Beacock & Tannett, Shepherd, Hill & Co., Thomas Jennings
       and others making machine tools.
,,   First jute machinery sent by Fairbairn's to New York.
1851. The Great Exhibition of all Nations, in Hyde Park, London.
,,   Building of Leeds Town Hall commenced.
,,   Isaac Singer invented his sewing machine.
1854. The Crimean War brought much work to Leeds firms.
,,   Richard Peacock joined Charles Beyer to start Beyer, Peacock & Co.
1855. John Fowler patented method of ploughing with aid of steam engines.
,,   Henry Bessemer invented pneumatic process for making steel.
,,   Greenwood & Batley left Fairbairn's and started in Armley.
,,   Leeds firms begin exporting jute machinery to India.
1858. Sir Peter Fairbairn, engineer, Mayor of Leeds.
,,   Manning, Wardle & Co. started in Hunslet.
,,   British Association in Leeds, Prof. Richard Owen, President.
,,   Carrick, Marshall & Co. started Sun Foundry, later taken over by
       Hathorn, Davis & Davey, specialists in heavy pumps.
1859. Greenwood & Batley made band knife for cutting layers of cloth.
,,   Bessemer erected his own works in Sheffield.

1859. John Barran started wholesale clothing business in Leeds.
 ,, John Fowler made first steam ploughing tackle in Hunslet.
1860. Yorkshire Council of Education formed.
 ,, Free Trade policy completed by W. E. Gladstone.
1861. Hudswell, Clarke & Co. started in Hunslet.
 ,, American Civil War and cotton famine in Lancashire.
1862. Smith, Beacock & Tannett named their works The Victory Foundry.
1864. Hunslet Engine Co. started to make locomotives.
 ,, Profit sharing started by Henry Briggs & Son, of Whitwood.
1871. Leeds Tramway Co. started horse cars.
1875. Yorkshire College of Science founded.

The Great Exhibitions of 1851 and 1862 marked a very active period of creative engineering work. The importance of organised teaching of scientific and engineering subjects also became widely recognised, and the Science and Art Department of South Kensington was founded.

In 1860, J. J. E. Lenoir, and in 1865 Pierre Hugen, of France, made workable gas engines, and in 1876 E. Langen and Otto, of Germany, invented their four cycle compression gas engine, which was afterwards made by Crossleys. The exploitation of mineral oil in U.S.A. caused oil engines to come into use.

The invention of the separately excited dynamo by Henry Wylde in 1863 and the work of C. W. Siemens and Z. T. Gramme in 1867 and 1870 in making workable dynamo machines brought new possibilities. Dynamos are essentially quick speed machines, and in due course there was a demand for engines to run at high speeds to couple direct to dynamos.

Amongst inventive engineers who began to work out ways of meeting this demand there were the Hon. Charles A. Parsons, who was then with Kitson's, of Hunslet, and Peter W. Willans, a connection of the Leeds cloth manufacturing family of that name.

The central valve engine of Willans was a great success, and partly with the collaboration of Col. R. E. Crompton, a Yorkshire electrical engineer, with works in Chelmsford, the engine set the pace for many years in electric light stations.

Sir Charles A. Parsons then turned his genius to inventing the steam turbine, and left Leeds to join Clarke, Chapman & Co., of Gateshead-on-Tyne, where his first turbine was made in 1884. Leeds, therefore, just missed the honour.

In five years the firm made over 300 turbo-generators, ranging in capacity up to 75 kilowatts. Sir Charles Parsons then started his own works, and in 1892 built the first condensing turbine, which marked a great advance in efficiency for the steam consumption per kilowatt hour or unit of electricity was reduced about 50 per cent.

By means of further improvements and the enterprise of several engineers on the North-East Coast, the Parsons' steam turbine came into use for propelling vessels, the " Turbinia " of 1897 being the first example.

In the meantime water-tube boilers came into general use for electricity stations because of rapid steaming and the possibility of building them for much greater outputs and pressures than was possible with Lancashire boilers.

A Brush arc lighting plant was installed in the works of John Fowler & Co. Ltd. about 1880. The invention of the incandescent lamp by Sir J. Wilson Swan and T. A. Edison was followed by the discovery of the electric motor. At the first Paris electrical Exhibition the accidental mis-connection of a dynamo showed that it would work as a motor.

The first electricity works in Whitehall Road, Leeds, had Fowler slow-running engines and alternators. Extensions were made with Belliss & Morcom quick speed engines, and later they also were replaced by large turbo-alternators.

In 1891 a pioneer electric tramway was laid down in Roundhay Road. Electric traction is now the method for suburban railways and it is a coming system for main line traffic.

In recent years the Diesel engine, working with high compressions and using heavy oils, has made great headway, and to meet this competition, steam pressures have been increased to as much as 1,500 lbs. per sq. inch and the super-heating of steam to over 700 deg. Fah.

The latest development in boiler practice is the burning of coal in powdered form, as a mechanical gas. Efficiencies approach 90 per cent. and modern boilers can be made for outputs forty times as great as that of a Lancashire type.

INDUSTRIAL CO-PARTNERSHIP.

In these days when so much is heard about the advantages of profit-sharing it is of interest to know that Leeds was the birthplace of the idea in this country. Industrial co-partnership was evolved as early as 1865 by Henry Briggs & Sons, of Whitwood and Methley collieries, when they arranged for any of their workmen who cared to do so, to be partners.

Briefly, the arrangement was such that at the end of the year, an amount equal to 10 per cent. on capital invested was deducted from profits to cover depreciation and interest and the balance of profits were divided into equal parts, one being divided amongst those workpeople who were partners. The system was carried on for several years and even after it stopped the workpeople were still allowed to have a director.

It is possible that the idea came to Henry Briggs from the " Round Foundry," Holbeck, because in 1844 a number of employees took over the business of Fenton, Murray & Jackson, and worked for some years on profit-sharing lines. They were numerous enough to be nicknamed the " Forty Thieves," no doubt from a pantomime of that name at the Theatre Royal, Hunslet.

Any who wanted to be paid out, received their money from the others who remained in, and this went on until half-a-dozen were left. They combined to form the well-known firm of Smith, Beacock & Tannett, and concentrated on machine tool making.

Soon after the experiment at Whitwood started, two firms of iron manufacturers, namely, Clayton Plate and Bar Iron Co. and Fox, Head & Co., of Middlesbrough, tried similar schemes.

In 1867 John and Henry Gwynne, engineers, of Hammersmith, London, commenced profit-sharing, this eing the first case of an engineering concern to do so. They were followed in 1870 by the North of England Industrial Coal and Iron Co., Middlesbrough, and by that time about fifteen firms had tried profit-sharing or co-partnership.

It should be mentioned that Thomas Hughes, the author of " Tom Brown's School Days," was a factor in causing Henry Briggs and others to pioneer in this direction. Unfortunately the trade slump that followed the Franco-German war caused discouragement, and only a few concerns carried on.

The idea of co-partnership of employees was brought to life again about 37 years ago by another Yorkshireman, George Livesay, who introduced it into the South Metropolitan Gas Company, of which he was manager.

The leaders of the Gas Workers' Trade Union objected because they thought they saw in it a weakening of their influence, and they called out the gas stokers. The strike failed because others were easily found to do the work and accept co-partnership.

It has been a remarkable success, and the Gas Light and Coke Co. and other concerns who also have the method in operation have found it a means of maintaining amicable and efficient industrial relations.

Messrs. Taylor, of Batley, and W. Mabane, of Leeds, are " carrying the torch " of industrial co-partnership in the West Riding.